The Challenge of Reframing Engineering Education

Dennis Sale

The Challenge of Reframing Engineering Education

Springer

Dennis Sale
Singapore Polytechnic
Singapore
Singapore

ISBN 978-981-10-1191-7 ISBN 978-981-4560-29-0 (eBook)
DOI 10.1007/978-981-4560-29-0
Springer Singapore Heidelberg New York Dordrecht London

Softcover reprint of the hardcover 1st edition 2014

Printed on acid-free paper

Springer is part of Springer Science+Business Media (www.springer.com)

Preface

> We understand everything in human life through stories.
> —Jean-Paul Sartre

This book is a story about improving the quality of the student learning experience at Singapore Polytechnic (SP), Singapore's first polytechnic. This was achieved through reframing a number of engineering diploma courses based on the Conceive-Design-Implement-Operate (CDIO) engineering education framework, originally conceived at the Massachusetts Institute of Technology (MIT).

SP, like many educational institutions globally, must tackle challenges of maintaining competitive advantage in the educational sphere which involves, among other things, identifying the most relevant competencies for the twenty-first century worker and citizen, designing a curriculum that offers meaningful and enriching experiences for its students and, inevitably, enhancing faculty competence in all things to do with teaching and learning. Now, we know that defining *the most relevant* curriculum outcomes, models and pedagogic arrangements is long contested and open both to valuations and future societal challenges—much of which is unknown. However, choices have to be made and SP decided to adopt the CDIO framework as it offered a robust curriculum development approach as well as the necessary flexibility for local customization and creative adaptation.

The book draws on the implementation experience from 2004 until the time of writing, and has the primary purpose of providing a comprehensive resource to help other institutions in a similar process of, or considering, large-scale curriculum reframing. It stemmed from the essential need to thoughtfully manage the many activities involved, such as tracking agreed outcomes (e.g. what is to be done, when and by who), the development and appraisal of curriculum artefacts (e.g. syllabus documents, learning designs/activities, resources, assessments, etc.) and not least, providing the range and types of faculty development programmes and support resources as and when needed. Invariably it dawned on me that our experience could provide practical value for others engaged in similar projects. Having completed a 3-year longitudinal evaluation of the student learning experience, in which students were active co-participants in the research process, enabling us to collaboratively learn what worked and how, I feel that the story is now worth telling—hence this published text.

In writing this book, I have been especially mindful of the following considerations. First, to make all chapter topics practically focused, enabling the reader to see clear applications in a range of contexts. Second, to ensure a strong evidence-base for all the approaches taken, referring to validated research findings and practitioner experience. Third, I confess to have deliberately woven into the narrative a more informal style than is typical of the genre. Being of Cockney descent, from East London, it is not surprising I like the quote by John Dewey, *To be playful and serious at the same time is possible, in fact it defines the ideal mental condition.* You can be sure; implementing large-scale curriculum innovation will be helped through such a mindset.

While the book has its primary focus on reframing engineering education, based on the CDIO engineering education framework, much of the content, especially an explicit model for developing good thinking and a comprehensive research-based pedagogic approach for creative learning design, applies equally well to all mainstream educational sectors and contexts. After all, it is not just engineers who need an effective and exciting learning experience.

I hope you find the book an interesting read, apart from being a practical and useful resource for your curriculum innovation projects.

Acknowledgments

Successful curriculum implementation, by its very nature, requires a high level of sustained collaborative activity. Hence, without the cooperation and friendship of many academic faculty from the various schools and departments involved (Architecture and the Built Environment; Chemical and Life Sciences; Communication, Arts and Social Sciences; Digital Media and Infocomm Technology; Electrical and Electronic Engineering; Mathematics and Science; Mechanical and Aeronautical Engineering), I would not have produced this book.

Invariably, while many have contributed, certain individuals played key roles in making the book happen. My initial thanks go to Pee Suat Hoon who introduced CDIO at Singapore Polytechnic and demonstrated through her everyday teaching practices how engineering could be a really interesting and meaningful experience for students. It was her foresight and willingness to take on the challenge of reframing her curriculum and pedagogic approaches that showed others what might be possible at institutional level. Cheah Sin Moh provided on-going reassurance over the years with his comprehensive and creative re-design of the Diploma in Chemical Engineering programme. In the process, we researched key aspects of the student learning experience, co-wrote a number of papers, and managed to keep enthusiastic over a period of some 5 years—which is a tough call. Helene Leong, for our many years of working together on all aspects of programme implementation, the lively discussions on all things to do with enhancing student learning and a massive desire to do it all properly. Thanks Helene for going the distance.

To the many other SP faculty members who have contributed over the years, I hope I have done some justice to your work in implementing CDIO at Singapore Polytechnic. In the process of creating, designing, operating and implementing solutions to our educational challenges, I have learned much about the practices of engineering. I now see it for the exciting discipline it really is, so I thank you for that also.

Contents

Abstract

The Challenge of Reframing Engineering Education is the first book to document the experience of implementing the CDIO Engineering Educational Framework in a large educational institution in the Asian context. It focuses on how to successfully implement and manage the key stages, activities and inevitable challenges that have to be negotiated in any large scale curriculum innovation. Its main purpose is to provide a practical resource for curriculum innovators and practitioners on what needs to be done, how and on what basis. It is written in a more narrative style than is typical of the genre, engaging the reader more intimately with the actual decision making processes and rationale that underpins curriculum innovation in the real context of institutional life. The book also encompasses many innovative practices for supporting student learning which are relevant in all mainstream educational contexts. These include an evidence-based learning approach for creative teaching, an explicit model for developing good thinking and a design framework for producing effective and efficient blended learning.

Keywords Engineering education · CDIO approach · Curriculum innovation · Creative learning design · Blended learning

Chapter 1
Introduction

1.1 Context

This book is based on the experience of Singapore Polytechnic (SP), one of the world's largest polytechnics, as it reframed a number of engineering education programmes based on the Conceive-Design-Implement-Operate (CDIO) engineering education framework.

The experience is unique in that SP was the first higher educational institution in Asia to implement CDIO and inevitably had to do its own reframing of an educational model conceived in a western university context to the Asian and polytechnic context of Singapore. Most significantly, in this process of reframing, SP developed many innovative and practical approaches to curriculum design, teaching and assessment, which have culminated in enhancing the learning experience for its students.

The book addresses the various activities and challenges that require careful negotiation in any large scale curriculum innovation, as well as offering examples and insight to help guide other educational institutions embarking on similar curriculum innovation projects. It is deliberately written in a more informal narrative style than is typical of the genre, though emphasizes a strong research base from which to validate good practice.

1.2 CDIO Framework and Rationale

CDIO is an international initiative, originally conceived at the Massachusetts Institute of Technology (MIT) in Cambridge, Massachusetts in the late 1990s. In 2000, in collaboration with the Swedish universities of the Wallenberg Foundation, Chalmers University of Technology, Linkoping University and the Royal Institute of Technology, the CDIO initiative was formed.

The initiative represented a response by engineers in industry, government and academia to a concern about the present state of engineering education.

D. Sale, *The Challenge of Reframing Engineering Education*,
DOI: 10.1007/978-981-4560-29-0_1, © Springer Science+Business Media Singapore 2014

Essentially, engineering education was seen as over prioritizing the teaching of theory, especially mathematics and science, while not paying enough attention to the real world of engineering practice and the need for skills such as design, teamwork and communications. As Crawley et al. (2007) summarized:

> …we identified an underlying critical need—to educate students who are able to Conceive-Design-Implement-Operate complex, value added engineering products, processes and systems in a modern, team-based environment. It is from this emphasis on the product, process, or system lifecycle that the initiative derives its name—CDIO. (p. 1)

The importance of Conceive-Design-Implement-Operate, as an organizing frame for engineering education is further explained by Crawley et al. (2007):

> Modern engineers lead or are involved in all phases of a product, process, or system lifecycle. That is, they Conceive, Design, Implement, and Operate. (p. 8)

The creation of an educational framework that would encompass this wide range of competencies, as well as the methodology for enabling the development of student capability to be able to conceive-design-implement-operate and be "ready to engineer" (Crawley et al. 2007, p. 6) is the basis and rationale of the CDIO engineering education framework.

All aspects of the CDIO engineering education framework are fully documented, explained and illustrated in 'Rethinking Engineering Education' (Crawley et al. 2007). The purpose of this book is not to replicate the what and how of CDIO at the generic level, but focus on the experience of SP in actually putting it into practice over the past 7 years—what has been learned from this endeavour and how it may be of practical use for other institutions who may be working towards similar educational goals. As the old saying goes, let's not "re-invent the wheel"—so to speak. The book aims to contribute to the learning and development of the CDIO community, as well as providing an invaluable resource to aid similar curriculum development activity in vocational and higher education, whether at university, polytechnic or technical college level contexts.

However, in order to provide context for the reader who may be unfamiliar with CDIO, certain key aspects of the framework will be briefly identified and explained in order to ensure easy transition through the main chapters of the book.

Firstly, one of the refreshing aspects of the CDIO approach is the non-allegiance to any one particular perspective in education or a lock step and highly prescriptive procedural approach to implementation. As noted by Crawley et al. (2007):

> Nothing in our approach is prescriptive. The CDIO approach must be adapted to each programme—its goals, university, national, and disciplinary contexts. It is aligned with many other movements for educational change, but unlike national accreditation and assessment standards that state objectives, we provide a pallet of potential solutions to the comprehensive reform of engineering education. (p. 4)

However, within this broad eclectic approach, which will be fully explicated and illustrated in the various chapters, there is a robust curriculum model and a sound pedagogic foundation encouraging active and experiential learning in solving real world engineering problems. Central to the curriculum approach is the specified CDIO learning outcomes, derived from extensive stakeholder collaboration and tested by peer review, which form the basis for program design and assessment. Crawley et al. (2007) summarizes as follows:

> A CDIO programme creates a curriculum organized around mutually supporting technical disciplines with personal and interpersonal skills, and product, process and system building skills highly interwoven. These programs are rich with student design-implement experiences conducted in modern workspaces. They feature active and experiential learning and are continuously improved through a robust, quality assessment process. (p. 3)

The implementation of a CDIO program is guided by 12 standards, which define the key features of the approach, establish benchmarks of good practice and serve as the basis for continual improvement. The 12 CDIO Standards address:

- Program philosophy (Standard 1)
- Curriculum development (Standards 2, 3 and 4)
- Design-implement experiences and workspaces (Standards 5 and 6)
- Methods of teaching and learning (Standards 7 and 8)
- Faculty development (Standards 9 and 10)
- Assessment and Evaluation (Standards 11 and 12)

Table 1.1 summarizes the 12 CDIO Standards (Crawley et al. 2007, p. 35). Subsequent chapters deal with the various standards in the context of actual curriculum development and implementation.

1.3 Chapters in this Book

Each chapter focuses on particular aspects of implementing curriculum reframing and innovation based on the CDIO approach. However, while chapters may focus on specific CDIO standards and how SP has implemented them in practice, there is incorporated a wider exploration of what constitutes good practice in relation to all aspects of curriculum development, teaching and assessment. The book, in CDIO tradition, customizes the approach to the context of the polytechnic environment in Singapore and seeks to further advance aspects of CDIO pedagogy and assessment practices based on our experience, research and attempts at programme improvement.

Chapter 2 'Producing Curriculum Outcomes' documents the challenge of customizing the general CDIO syllabus to one that was contextualized to the needs and proficiency level of polytechnic students in the Singapore context. It will demonstrate and illustrate the necessary processes and strategies that proved important in arriving at clear and appropriate learning outcomes, which are fundamental to the success of any curriculum initiative. The importance of open collaboration and persistence in ensuring that the resulting learning outcomes are

Table 1.1 Summary of CDIO standards

CDIO standard	Summary description
1. The context	Adoption of the principle that product, process, and system lifecycle development and deployment—Conceiving-Designing-Implementing-Operating—are the context for engineering education
2. Learning outcomes	Specific, detailed learning outcomes for personal and interpersonal skills; and product, process and system building skills, as well as disciplinary knowledge, consistent with program goals and validated by program stakeholders
3. Integrated curriculum	A curriculum designed with mutually supporting disciplinary course, with an explicit plan to integrate personal and interpersonal skills; and product, process, and system building skills
4. Introduction to engineering	An introductory course that provides the framework for engineering practice in product, process, and system building, and introduces essential personal and interpersonal skills
5. Design-implement experiences	A curriculum that includes two or more design-implement experiences, including one at a basic level and one at an advanced level
6. Engineering workspaces	Engineering workspaces and laboratories that support and encourage hands-on learning of product, process, and system building, disciplinary knowledge and social learning
7. Integrated learning experiences	Integrated learning experiences that lead to the acquisition of disciplinary knowledge, as well as personal and interpersonal skills; and product, process, and system building skills
8. Active learning	Teaching and learning based on active and experiential learning methods
9. Enhancement of faculty competence	Actions that enhance faculty competence in personal and interpersonal skills; and product, process, and system building skills
10. Enhancement of faculty teaching competence	Actions that enhance faculty competence in providing integrated learning experiences, in using active experiential learning methods, and in assessing student learning
11. Learning assessment	Assessment of student learning in personal and interpersonal skills; and product, process, and system building skills, as well as disciplinary knowledge
12. Program evaluation	A system that evaluates programs against these standards, and provides feedback to students, faculty, and other stakeholders for the purposes of continual improvement

most relevant and well constituted in terms of clarity and proficiency level is emphasized.

Chapter 3 'Effective and Creative Learning Design' builds upon the active and experiential learning focus advocated by CDIO, but extends the pedagogic approach into a more comprehensive learning framework, based on the most relevant knowledge relating to how humans learn. While active and experiential learning are foundational principles for effective learning, there are other important principles of learning that need to be incorporated into a fully *science of learning* pedagogic framework.

Chapter 4 ‘Assessing Learning’ fully explores the key principles, procedures and practices of good assessment. It highlights the need to apply established principles of good assessment in accordance with assessment standards for the key stages of the assessment process. Ways to enhance the effectiveness and efficiency of the assessment process are identified, explained and illustrated in the context of CDIO.

Chapter 5 ‘Using Information-Communication Technologies to Support Learning and Teaching’ is not a specific feature of CDIO as such. However, as information-communication technologies are now an integral part of teaching and learning, a consideration of their effective use merits inclusion. This is especially the case as much of current practice in the area of technology use and e-learning in the educational context has failed to utilize the affordances that technology can offer in terms of enhancing teaching and learning. The chapter offers an approach to using emerging technologies and, in particular, blended learning, to exploit specific technology affordances underpinned by sound pedagogic design.

Chapter 6 ‘Evaluating the CDIO Experience’ documents our 3 year evaluation of the implementation of CDIO at SP. The evaluation used a range of data sources for evaluating key aspects of the CDIO programme from both students and faculty perspective. A particularly interesting and valuable aspect of the evaluation methodology was the involvement of student co-participants in the research programme. They met every semester for informal panel interviews with the evaluation team, and blogged regularly on their (and their classmates) learning experiences in specific CDIO infused modules. This enabled us to gain a more ethnographically based and longitudinal understanding of how our students experienced various aspects of the curriculum innovation.

Chapter 7 ‘Managing the Change Process: Approach, Strategies and Professional Development’ summarizes our management of change approach and strategies as well as the evolvement of our professional development support activities. It provides a realistic approach to the management of a large scale curriculum innovation, emphasizing the importance of openness, relationship building and actually doing things *better* rather than just quicker.

Reference

Crawley E, Malmqvist J, Ostlund S, Brodeur D (2007) Rethinking engineering education. Springer, New York

Chapter 2
Producing Curriculum Outcomes

2.1 Introduction

> Curriculum can be seen as the battlefield of many competing influences and ideologies. (Kelly 1989 p. 149)

Defining curriculum outcomes is essentially concerned with addressing the question of what skills, knowledge and attitudes are most useful to attain and for what purpose. This is of course a contested issue as it inevitably reflects views about the 'good society' and what are the 'desirable' attributes' of people living in that society. The problem is illustrated by Wringe (1988), who wrote:

> Human beings have potential for developing in many directions and the problem of educational aims is deciding which kinds of development should be fostered and which discouraged. (p. 43)

In the specific context of engineering education, the issue of curriculum outcomes is captured by Crawley et al. (2007):

> What is the full set of knowledge, skills, and attitudes that engineering students should possess as they leave the university, and at what level of proficiency? (p. 34)

However, in engineering, as in most curriculum areas, we are being faced with an increasing major planning dilemma. On the one hand we are experiencing an exponential growth in subject content knowledge (Neff et al. 1995). They point out that apart from exponential knowledge growth, engineering is becoming increasingly specialized and changes rapidly. Simply tinkering with the content curriculum, adding, deleting, rationalizing, etc., cannot address this problem, and it can only get worse. At present, there is great pressure on many lecturers to cover more content in shorter time frames via compressed modules and the use of information-technology communication applications, such as online and blended learning delivery modes. On the other hand, there are expectations that graduates will not only have deep technical knowledge, but also a range of problem-solving skills, communication and teamwork competencies, and a disposition for flexible lifelong learning.

D. Sale, *The Challenge of Reframing Engineering Education*,

DOI: 10.1007/978-981-4560-29-0_2,

In the present societal context of rapid change, unpredictability and a volatile work context, we are swamped with knowledge possibilities and confusion about what are the most important human qualities and attributes to foster, both for the world of work and for effective citizenship. Quite simply, we cannot fully know what abilities or competencies will be most useful for future society. Curriculum planning must, therefore, recognize the systemic and accelerating nature of change and build the necessary flexibility into the curriculum format and learning process.

Educational systems, philosophy and practices inevitably reflect the societal context in which they prevail. They are also likely to incorporate the interests and concerns of dominant decision-making groups in that society. In Singapore, where this innovation is taking place, there is heavy reliance on the continuous development of its human resource in order to sustain and enhance competitive advantage. Such advantage will only be possible in future with a workforce capable of responding to the enormity and complexity of economic and technological change with both a high level of productiveness and creativity. Mr. Goh Chok Tong (1997), the former prime minister of Singapore once stated:

> The old formulae for success are unlikely to prepare our young for the new circumstances and new problems they will face. We do not even know what these problems will be, let alone be able to provide the answers and solutions to them. But we must ensure that our young can think for themselves, so that the next generation can find their own solutions to whatever new problems they may face. (p. 3)

The strong commitment to education in Singapore, a nation state renowned for systematic planning and efficiency is, therefore, hardly surprising. Again to quote Mr. Goh (1997) in this context:

> A nation's wealth in the Twenty-first century will depend on the capacity of its people to learn. Their imagination, their ability to seek out new technologies and ideas, and to apply them in everything they do will be the key source of economic growth. Their collective capacity to learn will determine the well-being of a nation. (p. 1)

However, while the context of Singapore has been highlighted here, such issues and concerns relating to educational aims and curriculum outcomes are generic to an increasing number of countries. Indeed, the scenario in the global context was perfectly captured by Reich (1992):

> In the emerging global economy, even the most impressive of positions in the most prestigious of organizations is vulnerable to worldwide competition…The only true competitive advantage lies in skill in solving, identifying and brokering new problems. (p. 148)

2.2 Challenges to Higher Education Institutions

As outlined in Chap. 1, the CDIO initiative was in large part a response to perceptions by stakeholders that present engineering courses are too theoretical and do little to generate interest let alone passion for engineering among students.

Furthermore, there are other widespread concerns raised concerning institutions of higher education, which further highlight the need for significant curriculum change. Diamond (1998), from a wide range of sources, argues that:

> A serious problem that institutions of higher education face is the perception by business leaders, governmental leaders, and the public at large that they have enthusiastically avoided stating clearly what competencies graduates should have and that as a result they have provided little evidence that they are successful at what they are expected to do. (p. 4)

In an increasingly competitive educational landscape, institutions of higher education must be seen as relevant by key stakeholders if they are to stay viable and thrive. It is not surprising that many are conducting major reviews and adopting what is referred to as 'Outcomes-Based Education' (OBE). OBE emphasizes that:

- The starting point for any curriculum offering is the identification of clear student learning outcomes, which describe the result of learning for that curriculum.
- The learning environment and instructional system is designed to promote the desired outcomes.
- Assessment supports the learning process (formative) and is more performance based (authentic assessment). Teaching, learning and assessment are systematically interlocked.

The CDIO approach has much in common with OBE and is consistent with a broad curriculum shift from 'content-based' to a 'competency' and 'process-based' focus. In basic terms, in the context of engineering education, there are increasing expectations that graduates will leave universities and colleges with both relevant practical competences as well as a thorough understanding of the role of an engineer in the present and future work context, and what this entails.

2.3 Deriving Specific CDIO Curriculum Outcomes

This section documents our approach to customizing the generic CDIO syllabus to the context of SP. It specifically focuses on addressing Standard 2: Learning Outcomes:

> Specific, detailed learning outcomes for personal and interpersonal skills; and product, process, and system building skills, as well as disciplinary knowledge, consistent with program goals and validated by program stakeholders.

This was a major challenge as we needed to produce a customized programme of curriculum outcomes that were authentic to the original CDIO skills framework but tailored to realistic range and proficiency levels for students in a polytechnic context. We were well aware that the result of this part of the curriculum reframing would largely determine the subsequent success (or otherwise) of developing and aligning the assessment approaches and instructional strategies.

The original CDIO Skills framework was the product of a comprehensive stakeholder focus group exercise comprised of engineering faculty, students, industry representatives, university review committees, alumni, and senior academicians. From this stakeholder exercise, it was agreed that every graduating engineer should be able to:

> Conceive-Design-Implement-Operate complex value-added engineering products, processes, and systems in a modern, team-based environment. (Crawley et al. p. 13)

To meet this aim a framework of learning outcomes was derived, which now constitutes the CDIO Syllabus. The syllabus classifies learning outcomes into four high-level categories:

1. Technical knowledge and reasoning.
2. Personal and professional skills and attributes.
3. Interpersonal skills: teamwork and communication.
4. Conceiving, designing, implementing, and operating systems in the enterprise and societal context.

These high level categories are further subdivided and organized into four discrete rational levels. While levels 1 and 2 are generic and specified, the selection of level 3 and 4 learning outcomes and the level of proficiency is within the framing of individual educational institutions, customized to the course context and stakeholder needs. The recommended process for establishing proficiency levels and learning outcomes is as follows:

- Review the generic CDIO Syllabus and make modifications or additions to customize it for a specific course of study within the technical and national context of the program.
- Identify and survey the important stakeholders of the program—both internal and external to the university—and validate their coverage and proficiency level to the local context.
- Write specific learning outcomes that guide the design of learning and define the assessment requirements.

This is a critical process for the success of the curriculum innovation. Limitations in the appropriateness, clarity and currency of the outcomes inevitably run through the instructional and assessment systems. There's limited value in teaching and assessing a knowledge or skill area in effective and efficient ways if it has little relevance to stakeholder interests. Furthermore, if the outcomes are not seen as clear and appropriate by faculty, there will be little buy in and the initiative will be bound for failure at an early stage. Diamond (1998) clearly recognized this when he argued that:

> …it is a major mistake to take any published list of basic skills or competencies and accept it for use on another campus without revision. Not only will the specific items on such a list vary from institution to institution but the definition of each item will vary as well. The final

> list of competencies, their definitions, and how they should be assessed must evolve on each campus. Faculty ownership in the process is an essential element for success. (p. 53)

In order to ensure that the CDIO skills at levels 3 and 4 were most appropriate to the context of students at SP a working group of representatives from the various engineering schools was established to systematically work through all the CDIO Skills, with a remit to:

- Identify which skills were most appropriate in the SP context.
- Decide a viable proficiency level.
- Write specific learning objectives that are measurable at level 4.

In practice this was a time-consuming process as faculty have different frames about what skills should be included, the level of proficiency deemed viable and the actual statements of specific learning outcomes. Our approach was to spend the necessary time and persevere in order to get the best possible consensus. While this resulted in a large number of meetings and iterations, in the longer run it was time well spent. The present SP customized syllabus is contained in Appendix 1.

2.4 Integrating CDIO Skills into Course and Module Curriculum

This specifically focuses on CDIO Standard 3: Integrated Curriculum:

> A curriculum designed with mutually supporting disciplinary courses, with an explicit plan to integrate personal and interpersonal skills; and product, process, and system building skills.

Having produced the customized CDIO syllabus, the next challenge was to naturally integrate specific skills within the curriculum programme and to ensure the best vertical and horizontal articulation possible (e.g., logical structuring and sequencing within and between modules in a course programme, etc.).

The notion and rationale for an integrated curriculum is not new. The nature of knowledge and its relationship to the development of mind has long been debated in the academic literature. The more traditional structuring of knowledge into disciplines and subjects, for example, is based on certain assumptions about the nature of knowledge, its relation to the development of mind and what it means to become educated. For example, Hirst (1974) argued that there is a close relationship between the acquisition of knowledge and the growth of the mind. Such an approach makes the following psychological assumptions about the nature of mind:

- Knowledge is a quality of mind. Failure to receive certain forms of knowledge is a failure to achieve rational ways of thinking in those areas
- The mind does not develop in a rational way—it needs organised forms of knowledge.

In contrast, Young (1971) argued that knowledge is less delineated at the experience level and is best learned in a more integrated and holistic context. In basic terms, a well-integrated curriculum is more consistent with how we learn, the nature of knowledge in the real world, as well as making learning more interesting for students (Fogarty 2009).

However, it is important to note that integrated approaches to curriculum do not negate the importance of subject domains and the importance of learning a core structure of knowledge that is foundational to understanding the discipline. The concern is that discipline knowledge is often overemphasized at the expense of knowledge connectedness. In the CDIO context, an integrated curriculum has the following important attributes:

- It is organized around the disciplines. However, the curriculum is re-tasked so that the disciplines are shown to be more connected and more supporting, in contrast to being separate and isolated.
- The personal and interpersonal skills, and product, process, and system building skills are highly interwoven into mutually supporting courses, relieving the potential tension between technical disciplines and these skills.
- Every course or learning experience sets specific learning outcomes in disciplinary knowledge, in personal and interpersonal skills, and in product, process, and system building skills, to ensure that students acquire the appropriate foundation for their futures as engineers (Crawley et al. 2007, p. 78).

The decision was taken to focus initially on Personal and Professional Skills and Attributes, and Interpersonal Skills: Teamwork and Communication for systematic integration into course curriculum. Firstly, it was felt that this would be more manageable and not lead to confusion and unrealistic workloads for faculty. Secondly, these specific skills were chosen for their relative familiarity with faculty and high leverage potential in the learning process. While many staff had not consciously thought about teaching thinking skills explicitly, the notion of integrating types of thinking with subject content knowledge was not a difficult selling point from a learning point of view. This was also the case for teamwork and communication. Good thinking, teamwork and communication are well established generic skills for effective learning and performance in real work contexts, whether in engineering or otherwise.

Each of the schools involved conducted a gap analysis of the courses selected for the integration of CDIO skills. From this, it was possible to:

- Identify where such skills were already apparent in the curriculum, whether explicitly stated or otherwise.
- Identify where there were naturally occurring opportunities to integrate selected CDIO skills.
- Map the integration of the CDIO skills throughout the course programme.

The mapping exercise, which required sustained collaboration and effort from the course team members, enabled the selected CDIO skills to be systematically infused, at the level of specific learning outcomes, into appropriate modules over the course duration.

Individual schools were encouraged to be creative in the way they integrated aspects of their course curriculum. For example, in the School of Mechanical and Aeronautical Engineering's *Introduction to Engineering,* knowledge and skills learned in two separate modules were integrated to conceive, design, and build a model racing car. Students machined the chassis from a given set of blueprints and applied creative thinking to conceive, design, and model the car's body. They then assembled and raced their cars. *Teamwork and communication* skills, as well as basic *Conceive, Design,* and *Implement* skills were introduced and woven into the activity. In year two, knowledge and skills taught in two existing modules, Engineering Design (ED) and Computer-Aided Design (CAD) were integrated to conceive, design and build a working industrial machine in the design-build module, illustrated in Fig. 2.1: Integrated Design Build Module. The integrated design-build module exposed students to the various stages of machine design such as conceptualization of design specifications, drawing, designing, fabrication, assembly, and commissioning of the machine.

Finally, it was recognized that while certain modules would focus on teaching and/or assessing specific skill components, all faculty would take responsibility for reinforcing such skills at teachable moments in their modules. In this way, we were seeking both a clear structure to the teaching and assessing of the skills, as well as encouraging an overall holistic and flexible approach to teaching beyond the mainstream technical content curriculum.

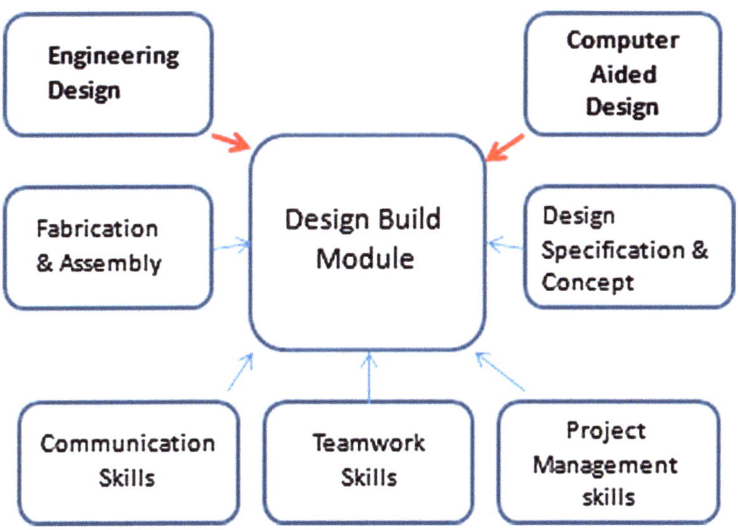

Fig. 2.1 Integrated Design Build Module

2.5 Using the Concept of Infusion for Integrating CDIO Skills

It is important that the CDIO skills are not simply 'put into' parts of the existing curriculum. Good integration is a bit like doing a jig-saw puzzle: the pieces all need to fit in their correct place to get the full intended picture. In reality, we may never complete the jig-saw puzzle, but it remains the operational goal—so to speak. Furthermore, in this context, there are decisions about how much *content* and *process* should be included in a curriculum programme. For example, on the one hand, there is virtual agreement among cognitive psychologists that effective thinking, however defined, needs an extensive and well organized knowledge base. As Resnick (1989) summarized:

Study after study shows that people who know more about a topic reason more profoundly about that topic than people who know little about it. (p. 4)

Equally, on the other hand, while thinking is developed through the acquisition and mental engagement with knowledge, knowledge is only made meaningful through thought. As Paul (1993) strongly argued:

> Thought is the key to knowledge. Knowledge is discovered by thinking, analyzed by thinking, organized by thinking, transformed by thinking, assessed by thinking, and, most importantly, *acquired* by thinking. (p. vii)

In working towards the best integration of the selected CDIO Skills into the technical content we modelled the *infusion* approach advocated by Swartz (1987) and the *nested* and *threaded* approaches documented by Fogarty (2009). The infusion approach argues that generic process skills such as thinking are best learned through "conceptual infusion" with the subject content. This involves identifying the ingredients of good thinking—"the skills, competencies, attitudes, dispositions, and activities of the good thinker"—and designing these into the structure of the lesson content (p. 125). The essential point is that good application of the thinking process and skills mutually develop the meaningful acquisition of knowledge to form understanding. Furthermore, specific types of thinking can be systematically developed in terms of level of proficiency and range of context application over the course duration.

Our approach, therefore, was to recognize the range of important components of effective learning and derive a pedagogically sound and viable structure for the infusion of CDIO Skills. In the specific case of Sect 2.4 'Personal Skills and Attitudes', for example, this has involved identifying where in the subject content exist the richest opportunities to infuse the desired thinking and learning-to-learn skills.

To do this in the most authentic way, we created partnerships in which engineering subject specialists worked with Education Advisors from the Department of Educational Development (EDU) and collaboratively unpacked the content of module syllabi. This process was invaluable in many positive ways for enhancing the curriculum. Firstly, in some modules, it was noticed that there existed redundant content, or areas that were only of tertiary value to the intended learning outcomes. This led to content reduction, rationalization, and updating in many modules. Secondly, in working through the learning outcomes for the module topics, it was apparent that many contained a preponderance of 'knowledge' and 'comprehension' based learning outcomes—based on Blooms Taxonomy (1956), which is used as the basis for writing learning outcomes in SP. Also, a number of learning outcomes were not clearly written (e.g., discuss, develop, etc.). In response, we re-focused the key learning more towards real world engineering application and subsequently wrote specific learning outcomes in clear performance terms. Thirdly, it helped faculty to develop what Schulman (1991) referred to as 'Pedagogic Content Knowledge'. Essentially, this involves a deep understanding of the structure of specific topic areas in a domain, and how they are best taught.

Once modules became more real world focused, it is then possible to use cognitive modelling of the key activities involved in order to identify the types of thinking that underpin highly effective performance. This is typically achieved by firstly asking the subject specialists (e.g., academic faculty) to make explicit their thinking in relation to the following question:

> How would a highly competent person think in the effective execution of this activity?

A useful technique to facilitate this is to visualize the activity and try to systematically describe the stages and types of thinking involved in conducting it effectively. For many lecturers this took some time initially and is most profitably done in a small group of similar subject specialists. The difficulty is that experts, in any field, usually take for granted much of what is involved as this becomes tacit and unconscious over time. It requires some skill in cognitive modelling, as well as persistence, to make such knowledge and thinking processes explicit and transferable.

However, once knowledge and the thinking processes are made explicit, it is then possible to identify the specific types of thinking that underpins competence in the performance/activity under review. For example, in one module a significant topic area is 'Managing Pollution'. From this process the following types of thinking were identified as important in competent task performance and written as contextualized specific learning outcomes:

- Compare and contrast pollution in a range of contexts.
- Analyse the basis of pollutants.
- Make inferences and interpretations concerning the causes of pollution in different situations.
- Generate possibilities in terms of managing/reducing pollutants.
- Evaluate pollution policies.

Having identified the main types of thinking within the specific subject domains, it seemed particularly useful to establish a framework for the explicit teaching of good thinking, which could be used by faculty as a basis from which they could integrate types of thinking into their content curriculum in a more systematic manner. One of the most striking and consistent findings from the evaluation (documented in Chap. 6 : Evaluating the CDIO Experience) was that faculty, in the main, were not explicitly teaching thinking skills, as evidenced by both student and faculty feedback, leading to confusion among many students as to what constitutes 'good thinking'.

Indeed, simply recognizing that certain internal cognitive processes—'thinking'—are particularly important in developing understanding, and the subsequent application of acquired knowledge, does little in itself to aid systematic development of such capability in our students. Without valid and practical definition of what constitutes such terms as 'critical thinking', 'creative thinking'—indeed, 'good thinking'—teaching faculty will find difficulty in teaching and assessing these desirable cognitive skills. Indeed, according to Wagner (2010), a lack of clarity concerning such skills is still prevalent globally:

> In schools, critical thinking has long been a buzz phrase. Educators pay lip service to its importance, but few can tell me what they mean by the phrase or how they teach and test it… (p. 16)

> For the most part, teachers haven't been trained to teach students how to think. (p. xxiv)

There is no shortage of models of thinking or lists of thinking skills, processes and dispositions (e.g., Marzano et al. 1988; Swartz and Parks 1994; Perkins 1994). Similarly, there seems to be a reasonable agreement that competence in 'thinking' can be developed through appropriate pedagogic strategies. How we have learned to think will determine in large part how we think, much the same as for any kind of learned activity. As Perkins (1995) points out "People can learn to think and act intelligently" (p. 18). Paul (1993) provides an interesting analogy between the development of mind and physical fitness. He points out that the mind, like the body, "has its own form of fitness or excellence" which is "caused by and reflected in activities done in accordance with standards (critically)" (p. 103). He goes on to argue that:

> A fit mind can successfully engage in the designing, fashioning, formulating, originating, or producing of intellectual products worthy of its challenging ends… Minds indifferent to standards and disciplined judgment tend to judge inexactly, inaccurately, inappropriately, prejudicially. (pp. 103–4)

2.6 An Explicit Model of Good Thinking

The model of thinking outlined here (Sale 2011) does not profess to capture all aspects of this elusive cognitive capability. Accurate conceptualization of internal cognitive processes is inherently problematic and invariably unreliable, especially across subject domains. However, we feel that it is sufficiently valid in terms of classifying the main types of thinking and the cognitive heuristics involved, thus enabling practical curriculum planning, teaching and assessment of good thinking. Furthermore, research suggests that while there is variation in how humans experience phenomena in the world—based on prior experience and selective perception, etc.—our common human apparatus and need orientation typically results in shared ways of experiencing the world. Indeed, without this commonality, the inter-subjectivity of everyday life would be even more problematic than it is already. For example, Marton (1981) pointed out that:

> …we have repeatedly found that phenomena, aspects of reality, are experienced (or conceptualized) in a relatively limited number of qualitatively different ways. (p. 181)

What this means is that while psychologists may solve problems is some qualitatively differently ways from engineers, both at the individual and collective level, there is much of similarity in the types of cognitive activity involved. For example, both will analyse situations (cases) looking for causation and areas of possible relatedness, make comparisons and contrast with similar cases, build up inferences and interpretations from ongoing perceptions and data accumulation, generate possible solutions and decide action based on chosen criteria. Around this swirl of cognitive activity, there will be an overall monitoring of what is going on—typically referred to as metacognition. Metacognition refers to the distinctively human capability of having awareness of, and the ability to monitor and control one's cognitive processing, as well as emotional states, in order to enhance learning and performance. It operates at both conscious and sub/unconscious levels.

The explicit model of thinking used in our CDIO implementation depicts six main types of thinking as shown in Fig. 2.2.

Table 2.1 summarizes the key heuristics that underlie these broad classification frames on different types of thinking.

In this model, analysis, compare and contrast, inference and interpretation and evaluation are typically employed during critical thinking; whereas generating possibilities, as the term implies, is predominantly in creative thinking.

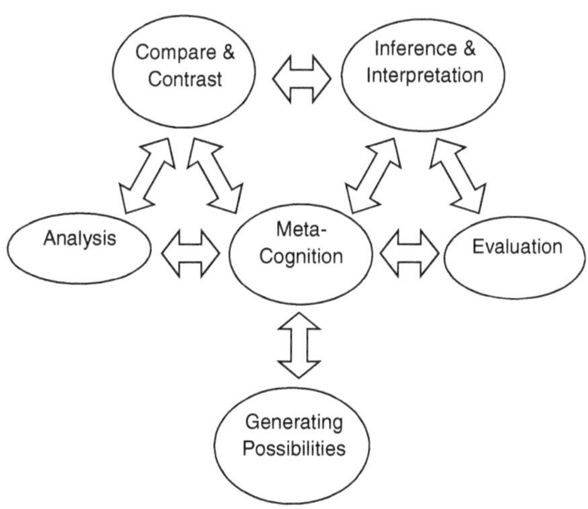

Fig. 2.2 Sale Model of Types of Thinking

Table 2.1 Summary of key heuristics of types of thinking

Generating Possibilities
• Generate many possibilities
• Generate different types of possibilities
• Generate novel possibilities
Compare and Contrast
• Identify what is similar between things—objects/options/ideas, etc.
• Identify what is different between things
• Identify and consider what is important about both the similarities and differences
• Identify a range of situations when the different features are applicable
Analysis
• Identify relationship of the parts to a whole in system/structure/model
• Identify functions of each part
• Identify consequences to the whole, if a part was missing
• Identify what collections of parts form important sub-systems of the whole
• Identify if and how certain parts have a synergetic effect (in an open system)
Inference and Interpretation
• Identify intentions and assumptions in data
• Separate fact from opinion in data
• Identify key points, connections, and contradictions in data
• Make meaning of the data/information available
• Establish a best picture to make predictions
Evaluation
• Decide on what is to be evaluated
• Identify appropriate criteria from which evaluation can be made
• Prioritize the importance of the criteria
• Apply the criteria and make decision
Meta-Cognition
• Recognize the ability to think in an organized manner
• Identify barriers (and limitations) to effective thinking
• Evaluate the effectiveness of thinking
• Identify strategies to enhance the quality of the thinking process

Metacognition, as outlined above, is the monitoring and evaluation of the overall thinking process (e.g. our execution of types of thinking), as well as other psychological aspects of our person that effect self-regulation (e.g. emotions, beliefs, etc.). In practice, these types of thinking run as overlapping and interdependent programmes, moving from foreground to background as the focus of a problem changes and certain questions emerge. Certainly, when creativity is sought, generating possibilities is at the mind's forefront, though other types of thinking will weave in and out of consciousness and, probably run continuously in the subconscious mind.

It is important to fully recognize that thinking processes do not exist in an 'uncontaminated' system of rational cognitions, but within the wider system of the human mind, composed of many interacting and conflicting parts or mental modules. As Pinker (2002), from extensive research, summarizes:

> Behaviour…comes from an internal struggle among mental modules with differing agendas and goals. (p. 40)

Marcus (2009), from a cognitive neuroscience perspective, fully highlights the challenge of achieving good critical thinking when he asserts that:

> Our beliefs are contaminated by the tricks of memory, by emotion, and by the vagaries of a perceptual system that really ought to be fully separate—not to mention a logic and inference system that is as yet, in the early twenty-first century, far from fully hatched. (p. 67)

Good thinking then is the ability to navigate this "perpetual swirl", and be able to employ the various heuristics of these types of thinking in a fluid, efficient and highly synergetic manner. This is perhaps the reason that good thinking is quite rare in many situations, and why we really need to teach it to our students.

It is in this context that some writers in the field see critical thinking not just in terms of cognitive processes and heuristics but also in terms of the development of intellectual traits and standards. For example, Paul et al. (2006) identify the following traits as central to acquiring a high level of expertise in critical thinking:

- Intellectual humility—sensitivity to owns own biases and the limitations of knowing.
- Intellectual courage—prepared to question own beliefs and those of others, even if unpopular with dominant perspectives and people.
- Intellectual empathy—awareness of need to actively entertain different views from one's own.
- Intellectual integrity—holding oneself to the same intellectual standards of others (no double standards).
- Intellectual perseverance—working through intellectual complexities despite frustration.
- Confidence in reason—recognizing that humankind's interests are best served by giving free play to reason.
- Intellectual autonomy—thinking for oneself in relation to standards of rationality and not uncritically accepting the judgements of others.
- Fair-mindedness—conscious of the need to treat all viewpoints alike and not be influenced by vested interests.

Such dispositions are certainly desirable, but the extent to which some are more integral to deep seated personality traits is open to question, as is their successful development in a pedagogic context. However, they remain a regulatory ideal and as educationalists we do our best to encourage productive outcomes for our students.

In summary, our approach to identifying good thinking started from the initial cognitive modelling of what highly competent professionals do when they solve real world problems, in order to derive domain contextualized thinking skills. This became the basis for identifying the types of thinking skills that naturally supported proficiency in the subject knowledge and skill areas. From this process we were able to provide systematic guidance in helping faculty to:

1. Review existing module aims and learning objectives to identify real world activities that students would be expected to do as a result of successfully completing the module.
2. Identify the types of thinking essential for highly competent performance in these real world activities.
3. Write learning outcomes that specifically cued the types of thinking in relation to knowledge acquisition.

The generic model of good thinking outlined in this chapter is helping to create a common language relating to thinking. It was very apparent from evaluation that both students and staff have wide variation in terms of their perception on what thinking entails and the cognitive heuristics involved. The model is already providing a more systematic approach to promoting good thinking in the curriculum, both in terms of the explicit teaching of thinking to students and staff development in this area.

Care was taken to ensure development of the skill components over time and ensuring sufficient practice for the attainment of understanding and competence as defined by the proficiency level. For example, as critical thinking was to be developed over the duration of the course programme, it is essential to have the structured development of such sub-skills as analysis, comparison and contrast, inference and interpretation, evaluation, over the duration of the course programme. As Marzano et al. (1988) pointed out:

> … we can improve students' ability to perform the various processes by increasing their awareness of the component skills and by increasing their skill proficiency through conscious practice. (p. 65)

This approach also proved useful for integrating other CDIO skills. We wanted to ensure that CDIO skills were not only appropriately integrated but actually supported and added value to the learning of subject content knowledge in the modules. As individual schools were at liberty to customize objectives at level 4 to the specific engineering context where appropriate, providing there is no change in the knowledge domain covered, cognitive activity involved and proficiency level, they could fully contextualize the CDIO skills. For example, in one of the engineering modules, specific learning outcomes for communication and teamwork were customized to the module content in the following ways:

- Apply team ground-rules and display teamwork (including leadership) in a range of team role situations when conducting experiments.
- Design appropriate communication strategies for presenting experimental findings.
- Demonstrate effective oral communication in presenting experimental findings.

The process of curriculum review for the implementation of CDIO led to a critical holistic re-framing of aspects of existing course structures and module integration, especially the re-writing and rationalization of specific learning outcomes. This is illustrated in Table 2.2 which shows a sample of learning outcomes before and after the curriculum review for a Chemical Reaction Engineering module.

Table 2.2 Sample learning outcomes before and after curriculum review in a chemical reaction engineering module

Before
• Distinguish between elementary and non-elementary reactions
• Explain the rate law and rate constant for elementary reactions
• Describe the temperature dependence of the rate constant using Arrhenius Equation
• Explain the molecularity and order of reaction
• Discuss the factors affecting the rate of reaction
• Determine the frequency factor and activation energy of a reaction
• Describe the steps involved for determining the rate law parameters
After
• Use Arrhenius Law to determine the effect of temperature on the rate of chemical reactions
• Infer and interpret experimental data on the effect of temperature on the rate of chemical reactions
• Compare and contrast the integral and differential methods of analysis in rate law determination
• Use integral and differentiated methods of analysis to determine the rate law for a liquid reaction
• Calculate and interpret the results for the integral and differential methods of analysis using graphical solution and linear regression
• Identify the components of an effective team
• Identify team roles and their impact on team performance
• Deliver effective oral communication to a given audience

2.7 The Importance of Identifying Key Underpinning Knowledge for CDIO Skills

We were well aware that teaching and assessing certain CDIO skills may pose significant challenges for some engineering faculty. Firstly, they may question the rationale for teaching such skills as contained within Personal and Professional Skills and Attributes, as well as Interpersonal Skills: Teamwork and Communication. Most significantly, is it our responsibility to teach these skills anyway? After all, we already have certain institutional modules and electives that cover many of these skill areas. Furthermore, are we not already overburdened with the demands of the technical engineering curriculum—not to mention the increasing plethora of administrative work that is becoming the norm in many educational institutions?

Secondly, and equally important, are faculty fully equipped to teach these skills effectively and efficiently? Many CDIO skills involve key knowledge from the fields of psychology, economics and, some would say, philosophy. The issue of whether all engineering faculty should be capable of teaching all CDIO Skills is significant. Similarly, a related issue concerns whether these are best taught by engineering faculty or more specialized faculty from the respective fields. For example, in the case of 'communication and teamwork', is it preferable and more viable for a specialist servicing department to take the main responsibility for teaching these skills rather than for engineering faculty?

However, what has proved helpful is the identification and delineation of the key underpinning knowledge for each of the CDIO skills. Underpinning knowledge refers to the key concepts, principles and procedures that are essential for developing understanding of the CDIO skill areas. This provides a conceptual basis for the practical applications to develop necessary competence, as illustrated in Table 2.3.

The production of clear and concise underpinning knowledge was important to facilitating 'buy in' from faculty. Once they fully understand what is involved and the importance of these skills for student learning, they were less resistant to the idea that they might need to teach such skills within the engineering context. Most significantly, they quickly appreciate that much of the underpinning knowledge—especially in the area of teamwork and communication—is, in fact, quite familiar to them. This is not surprising, as we would expect experienced engineering professionals to possess such knowledge and related competence. However, such knowledge is typically tacit (Polanyi 1967) rather than explicit. Through the provision of key underpinning knowledge for CDIO Skills, it is possible to bring such tacit knowledge to a more explicit and practical focus. Faculty can then see that they actually possess such knowledge and competence. It is then much easier for them to make direct connections to where and when in the curriculum such skills can be naturally and effectively integrated.

Key underpinning knowledge provided part of the solution to the question of whether certain CDIO skills were to be taught by engineering faculty or servicing departments who had specialist staff in these areas. At present, for communication

Table 2.3 CDIO skill and underpinning knowledge for managing learning

CDIO skill: Managing Learning

- Identify one's own learning approach
- Identify approaches for self-improvement (e.g., lifelong learning, creating positive beliefs and psychological states)
- Display key dispositions (e.g., initiative, perseverance, flexibility in work projects)
- Use a range of learning strategies and skills (e.g., goal setting, learning plans, organizing/summarizing information, receiving feedback)
- Managing time and resources

Underpinning knowledge for Managing Learning

- What is a learning approach and how does it impact on personal learning? Typical differences in the way people approach their learning (e.g., visual, auditory, kinaesthetic; holistic, serialist)
- Different ways in which self-improvement can be achieved (e.g., lifelong learning, creating positive beliefs and psychological states)
- The challenges that lifelong learning entails and its implications (continual re-skilling, job changes, professional and personal flexibility)
- What are positive beliefs and psychological states? How these can be developed and maintained (e.g., reframing, visualizing, self-motivation)
- What is meant by dispositions and how they impact on human behaviour? How certain dispositions (e.g., initiative, perseverance, flexibility) contribute to high performance and success in work projects, and in life goals
- What is meant by 'learning strategies and skills' and how they can help to make learning more effective and efficient? Different types of learning strategies and skills and how they contribute to improved learning (e.g., goal setting, learning plans, monitoring learning, organizing/summarizing information, receiving feedback)

and teamwork, a specialist module has been developed to cover the full range of skills as defined by the contextualized proficiency level. In terms of the actual teaching this is presently conducted by specialist faculty from the school of Communication, Arts and Social Sciences, working collaboratively with engineering faculty. Some engineering faculty are already sufficiently comfortable to integrate these skills into the module content curriculum and teach them as part of the mainstream engineering curriculum. Increasing faculty competence in taking on the challenges of teaching CDIO skills has been the result of the thorough preparatory work in writing clear and customized learning outcomes and familiarization with the supporting underpinning knowledge.

2.8 Summary

This chapter has documented and illustrated the process of customizing curriculum outcomes to a particular educational and institutional context. In conducting the range of activities involved, a number of related curriculum issues emerged which, in turn, required thoughtful consideration and action.

Firstly, while the process of customizing the curriculum outcomes was a time-consuming process, it was time well spent. It secured faculty ownership of the completed SP customized CDIO syllabus and ensured clarity and agreement on the range and proficiency of the curriculum outcomes. This was central to the successful implementation of the initiative.

Secondly, the process of integrating CDIO skills led to a more critical review of other aspects of course and module structure. As faculty had to review their module documents, it became apparent in a number of cases, that there was a need to modify the content organization as well as write clearer and more appropriate learning outcomes generally. Clear outcomes are central to curriculum alignment. Lack of clarity for this curriculum component typically creates confusion when it comes to the design of learning and assessment systems.

Finally, the decision taken to focus on a limited number of CDIO skills for initial integration proved a wise one as did the choice of the initial skills of Personal and Professional Skills and Attributes, and Interpersonal Skills: Teamwork and Communication. This reduced both the cognitive and work loads of academic faculty.

References

Bloom B (1956) Taxonomy of educational objectives. Longman, London

Crawley E, Malmqvist J, Ostlund S, Brodeur D (2007) Rethinking engineering education. Springer, New York

Diamond RM (1998) Designing & assessing courses & curricula: A Practical Guide. Jossey-Bass, San Francisco

Fogarty R (2009) How to integrate the curricula. Sage, London
Goh CT (1997) Ministry of Education Press Release No: 015/97 (28 April)
Goh CT (1997) Speech by prime minister Goh Chook Tong at the opening of The 7th international conference on thinking on Monday, 2 June at the Suntec City Convention Centre, Singapore
Hirst P (1974) Knowledge and the curriculum. Routeledge & Kegan Paul, London
Kelly AV (1989) The curriculum: theory and practice. Paul Chapman, London
Marcus G (2009) Kluge: the haphazard construction of the human mind. Faber & Faber, London
Marton F (1981) Phenomenography: describing conceptions of the world around us. Instr Sci 10:177–200.
Marzano RJ et al (1988) Dimensions of thinking: A framework for curriculum and instruction. ASCD, Alexandria, VA
Neff G et al (1995) Transforming engineering education from a product to a process. A paper presented at the 4th world conference on engineering education, 15–20 October, vol 1. Saint Paul, MN, pp 5–9
Paul RW (1993) Critical thinking. Foundation for Critical Thinking, Santa Rosa, CA
Paul R, Niewoehner R, Elder L (2006) The thinker's guide to engineering reasoning. Foundation for Critical Thinking, Tomales, CA
Polanyi M (1967) The tacit dimension. Routeledge & Kegan Paul, London
Reich R (1992) The work of nations. Vintage Books, New York
Resnick LB (1989) Assessing the thinking curriculum: new tools for educational reform. In: Gifford BR, O'Connor MC (eds) Future Assessments: Changing views of aptitude, achievement, and instruction. Kluwer Academic Publishers, Boston
Perkins DN (1994) What Creative Thinking Is: Excerpt from Perkins, D. N. creativity by design. Educ Leadersh vol 42(1):18–22
Perkins DN (1995) Outsmarting IQ: The emerging science of learnable intelligence. The Free Press, London
Sale D (2011) Writing learning outcomes for specific types of thinking. Paper presentation at international symposium on advances in technology education (ISATE) conference. 27–29 September 2011
Schulman L (1991) Pedagogic ways of knowing. Institute of Education, London
Swartz R J (1987) Teaching for thinking: a developmental model for the infusion of thinking skills into mainstream instruction. In: Baron JB and Sternberg RJ (eds) Teaching thinking skills: theory and practice. Freeman, New York
Swartz RJ, Parks S (1994) Infusing critical and creative thinking into content instruction. Critical Thinking Press & Software, Pacific Grove, CA
Wagner T (2010) The global achievement gap. Basic Books, London
Wringe C (1988) Understanding educational aims. Allen and Unwin, London
Young MFD et al (1971) Knowledge and control. Collier-Macmillan, London

Chapter 3
Effective and Creative Learning Design

3.1 Introduction

> ...the central problem of curriculum study is the gap between our ideals
>
> and our attempt to operationalize them. (Stenhouse 1989, p. 3)

The process of developing learning objectives appropriately calibrated to CDIO Standard 2: Learning Outcomes and customized to the context of Singapore Polytechnic was documented in the previous chapter. This chapter presents our approaches to teaching and learning and how we seek to address the essential question raised by Edstrom et al. (2007):

> How can we do better at ensuring that students learn these skills? (p. 130)

We were particularly diligent in the process of customizing the learning outcomes to the polytechnic context to ensure clarity and relevance. However, we fully appreciate that a nicely produced curriculum document does not guarantee quality student learning outcomes. Quality of education is, at the end of the day, predominantly a teaching quality issue. As Izumi and Evers (2002), from reviewing extensive research, concluded:

> ...nothing is as important to learning as the quality of a student's teacher. The difference between a good teacher and a bad teacher is so great that fifth-grade students who have poor teachers in grades three to five score roughly 50 *percentile points* below similar groups of students who are fortunate enough to have effective teachers. (ix)

Similarly, Rivers and Sanders (2002) point out:

> The effect of the teacher far overshadows classroom variables, such as previous achievement level of students, class size as it is currently operationalized, heterogeneity of students, and the ethnic and socioeconomic makeup of the classroom. (p. 17)

However, while we can certainly identify teaching quality as fundamental to the quality of student learning outcomes—and this would largely hold true irrespective of context (e.g., university, polytechnic, mainstream compulsory sectors, etc.)

D. Sale, *The Challenge of Reframing Engineering Education*,
DOI: 10.1007/978-981-4560-29-0_3,

—what constitutes highly effective teaching or 'good pedagogy' have long been contested questions in the educational literature (Tuckman 1995; Ornstein 1995; Darling-Hammond and Bransford 2005).

This chapter presents a pedagogic framework from which teaching and learning strategies can be effectively and creatively designed and delivered. It relates particularly to CDIO Standards 5: Design Implement Experiences; 6: Engineering Workplaces; 7: Integrated Learning experiences; and 8: Active Learning. Also, the chapter introduces a comprehensive *science of learning* approach that transcends the earlier CDIO constructivist perspective outlined by Crawley et al. (2007). It provides a pedagogic framework from which faculty can develop the range of learning design skills and practical teaching strategies that result in more engaging and effective student learning experiences, irrespective of delivery mode (e.g., face-to-face, online, blended).

3.2 The Pedagogic Foundation of CDIO

In broad terms this has been documented elsewhere. For example Crawley et al. (2007) stated:

> ...we recommend improvement in two basic areas: (1) an increase in *active and experiential learning*, and (2) the creation of *integrated learning experiences* that lead to the acquisition of both disciplinary knowledge, personal and interpersonal skills; and product, process and system building skills. (p. 29)

Active and experiential learning are not new pedagogic approaches. They have been extensively and successfully employed in a wide range of teaching and training contexts for many years. Essentially, active learning happens when students are given the opportunity to take a more interactive relationship with the subject matter of a course, encouraging them to generate rather than simply to receive knowledge. As Chickering and Gamson (1987) pointed out:

> Learning is not a spectator sport. Students do not learn much just by sitting in class listening to teachers, memorizing pre-packaged assignments, and spitting out answers. They must talk about what they are learning, write about it, relate it to past experiences, apply it to their daily lives. They must make what they learn part of themselves. (p. 3)

Active learning methods, when used effectively, engage students directly in the learning process, making possible the use of good thinking in relation to the key concepts, principles and procedures of subject disciplines. These are the very cognitive processes that enable learners to make meaning of their learning and build understanding, which makes possible the transfer of learning. The transfer of learning is as McTighe and Wiggins (2000) point out:

> ...our great and difficult mission because we need to put students in a position to learn far more, on their own, than they can ever learn from us. (p. 44)

Experiential learning is by its very nature is active, but emphasizes activities that more directly connect to the real world phenomenon being learned. For example, discussing the stages of effective team-formation is active learning, but is limited experientially. In contrast, actually participating in groups over time and experiencing the team processes first hand is both active and highly experiential.

However, it needs to be noted that experience is not a sufficient condition for effective learning. While we have long recognized that expertise in any field, and teaching is no exception, involves experience, we often fail to ask the question of what constitutes experience and what is its impact on learning and performance? Similarly, why is it that some people, who have many years of experience, still display limited competence, whereas relative newcomers achieve good competence in a comparatively short time? The conclusion of Berliner (1987) offers insight into such questions:

> …experience will probably only instruct those who have the motivation to excel in what they do and the metacognitive skills to learn from their experience…we believe that individuals with that kind of motivation to learn and in possession of a set of strategies for learning from experience are literally transformed by their experience. (p. 61)

There are some important points to emphasize in this context. Firstly, for experience to translate into meaningful learning there must be motivation on the part of the learner to critically unpack the experience and make the significant inferences and interpretations from it. This may prove a challenge to faculty who see their role primarily in terms of transmitting knowledge and conducting engineering based research. There needs to be a clear recognition that good teaching requires an increasing professional knowledge base in its own right as well as a range of integrated skill sets. Also, most importantly, there must be the motivation on the part of faculty to acquire such knowledge and apply it thoughtfully through a range of pedagogic practices.

Secondly, and equally important, the learner must have the requisite skills to negotiate this process effectively. For this reason, experiential learning methods are not simply concerned with the creation of experience per se, but the systematic assessment of such experience so that learners have the opportunity to assign meaning in relation to personal goals and expectations. This has implications both for faculty as they develop their own competence as teaching professionals, as well as their use of such methods with the students they teach.

The range and specific use of active and experiential methods are extensively documented in the literature and will not be reviewed here (Silberman 2005; Bonwell and Eison 1991). What is most important is how faculty, particularly those versed in more traditional didactic methods, approach requests to adopt more active and experiential methods of teaching and learning. Typically, some may feel initially daunted, which is a natural reaction to change, especially if the change is little understood and poorly supported. Supporting this change process is essential and will be discussed in some detail in Chap. 7.

The important point, however, is to recognize that there are many active and experiential learning methods, which involve varying types and levels of skill sets to utilize effectively. Using a more limited and comfortable range of such methods initially helps to build both competence and confidence for faculty. For example, posing good questions and using basic cooperative learning structures (e.g., Johnson et al. 1994) are not particularly difficult active learning methods, but are powerful when used appropriately and effectively. The highly effective use of Problem-Based Learning and Case-Based Learning involve a wider repertoire of skills—but again can be readily acquired by the enthusiastic teaching professional. After all, we are in the learning business and should be able to do this stuff—right?

There is little need to emphasize the point that the very nature of engineering is ideal for active and experiential learning methods. In basic terms, engineers solve problems—irrespective of who defines the problems for whatever purposes, etc. Active and experiential methods help to engage students in real world problem-solving.

The concept of integrated learning was outlined in Chap. 2 in relation to curriculum design, whereby connectivity is recognized both within and between disciplines. The concept of infusion was suggested as an approach to developing technical content knowledge and skills along with other skill sets such as types of thinking and communication. Such an approach fosters the kind of learning experience identified by Edstrom et al. (2007) when they argue that:

> Technical knowledge and the learning outcomes related to the CDIO Syllabus are interdependent and developed together... communication skills are deeply embedded in technical knowledge. (p. 134)

The infusion approach allows for the integration of diverse skill sets, providing they are naturally mutually supporting in the real world context. For example, it is more authentic to develop significant learning of ethical reasoning in the context of dealing with value-laden engineering scenarios. Einstein's dilemma in the Manhattan Project was an exemplar at the global level, but similar locally situated ethical issues face engineers on a regular basis, and their consequences are equally real to the folk affected.

3.3 A Science of Learning Framework: Developing Pedagogic Literacy

In the previous section, the pedagogic foundation to CDIO was identified and certain important aspects explored and contextualized. A key focus on incorporating more active and experiential methods into the teaching and learning approach was established as well constituted and necessary. However, recognizing that learning can be better fostered through these broad approaches does not address the full range of pedagogic considerations that underpin highly effective teaching, within the CDIO framework or in other curriculum contexts.

For example, while active learning can be highly effective in the ways documented, many teaching professionals often fall into the trap of providing 'activity for activity sake'. Unless the learning activities are clearly related to facilitating knowledge and skills contained in the learning outcomes, meaningful and appropriate to the particular learning group, sufficiently challenging but achievable, they are unlikely to get the desired results.

To design and deliver highly effective learning experiences across a range of learning contexts involves a much wider understanding of human learning and how this can be effectively applied to meet these ends. In the following sections the key considerations and knowledge bases will be outlined and illustrated. It will become apparent that the pedagogic framework presented aligns both to the CDIO approach, which aims to make engineering education more interesting and relevant to real world contexts, as well as other curriculum offerings with similar goals. Quite simply, good pedagogy applies across subject fields and educational sectors.

3.3.1 Defining Learning: What are the Important Issues?

There are many definitions of human learning in the literature. A popular enduring definition is that of Kimble (1961), who defined learning as:

> A relatively permanent change in behavioural potentialities that occurs as a result of reinforced practice. (p. 6)

Subsequent definitions, recognizing the multifaceted context in which learning occurs, highlight that learning is not adequately captured in such narrow psychological or behavioural terms. For example, Kolb (1995) suggests that:

> To learn is not the special province of a single specialized realm of human functioning such as cognition or perception. It involves the integrated functioning of the total organism—thinking, feeling, perceiving and behaving. (p. 148)

Similarly, Lave and Wenger (1993) argue that:

> …psychological theories of learning that conceive of learning as a special mental process, ultimately impoverish and misrecognize it. (p. 9)

However, from the standpoint of teaching, it is not so much the defining of learning that is important; rather an appreciation of the essential processes involved, how they work and the range of structural and situational factors that may promote or inhibit effective learning. From such understanding, teaching professionals can design and manage the range of learning environments and activities they offer learners in a more effective and efficient manner.

Furthermore, while learning is ultimately an individual act of personal construction and meaning making—as constructivists point out—there is real danger in 'over-individualizing' the learning process. The assertion by Schank (1997) is very pertinent:

> Contrary to common belief, people don't have different learning styles. They do, however, have different personalities. The distinction is important, because we need to be clear that everybody learns in the same way. (p. 48)

Coffield et al. (2004), in a comprehensive review, highlights the lack of reliable evidence that stable learning styles exist independently of the context in which they are expressed, and the lack of consensus about how teaching ought to be organized in light of these apparent differences. Similarly, Hattie (2009) from extensive meta-analysis of research correlating teaching approaches to student achievement concluded that "Emphasizing learning styles… are noted for their lack of impact (p. 199).

Indeed, the present vogue of highlighting differences in learners is potentially mitigating the more central issue of developing a strong body of empirical knowledge from which teaching and learning practices can be more validly and reliably enacted. Stone (2000) makes this point strongly when he comments:

> What teachers are told… is that student differences are important and if their teaching is truly creative, energetic and engaging, they will succeed in individualizing and bringing forth the best from all students. In effect teachers are being taught to make diagnoses that heighten their awareness of differences without advancing their ability to teach. (p. 43)

3.3.2 What do we Mean by Pedagogy?

Pedagogy, as with learning, has evoked much debate in terms of adequate definition. Approaches to pedagogy have gone through various phases, focusing on such aspects as 'teaching styles', 'paradigms of learning', 'models and methods of teaching' and the 'context of teaching'. As Mortimore (1999) points out

> Pedagogy has been seen by many within and outside the teaching profession as a somewhat vague concept. (p. 228)

Traditionally the term has been most used in relation to the teaching of children and has been contrasted with 'andragogy' (Knowles 1984), which argues that there are significant differences in the learning orientation of adults as compared to children, which necessitate different approaches to teaching and learning. More recent research and thinking relating to the pedagogy and andragogy debate suggest that while adults have certain different motivational bases to learning, as compared to children, and have different expectations about how their learning will be organized and managed, there are underpinning universal principles of learning, which have sufficient applicability across both learning groups.

Pedagogy then is not specifically focused on child or adult learning, or other factors such as gender, cultural or ethnic differences. It recognizes that how individuals orientate themselves to learning may differ in many ways, including the situated here and now context. To adopt an all encompassing and prescriptive pedagogic approach would be both limiting and potentially damaging. The approach considered most relevant for our purposes is captured by Mortimore (1999) who suggests that pedagogy is most usefully conceived as:

...any conscious activity by one person designed to enhance learning in another. (p. 3)

This conception of pedagogy also draws support from the work of Bain (2004) who documented how 'best teachers' were not constrained by particular approaches, methods or paradigms of learning. Instead they thought of teaching as:

...anything they might do to help and encourage students to learn. Teaching is engaging students, engineering an environment in which they learn. (p. 49)

3.3.3 Towards a Science of Learning

Sallis and Hingley (1991) made the observation that "education is a creature of fashion" (p. 9), which can influence the whole basis of curriculum orientation and teaching in particular. Furthermore, such *fashion* is largely driven by dominant paradigms in psychology or pedagogy. For example, the modern vogue is constructivism, which has gained favour over more behaviourist and transmission approaches. As a consequence, the emphasis is now on being student-centred and the role of the teacher has supposedly changed from being the 'sage on the stage' to the 'guide on the side'.

However, paradigms, by containing both premises and methodologies relating to particular domains of reality, limit both the effective and creative capability of working within that particular domain. The potential consequence of limiting practice largely to one dominant paradigm is well captured by Pratt (2002):

Perspectives are neither good or bad. They are simply philosophical orientations to knowledge, learning and the role and responsibility of being a teacher. Therefore, it is important to remember that each of these perspectives represents a legitimate view of teaching when enacted appropriately. Conversely, each holds the potential for poor teaching. (p. 14)

In this context then, while the present constructivist approach and 'active and experiential learning' focus for CDIO pedagogy represents an appropriate 'initial prototype' for framing teaching and learning, it is now important to enhance the design and subsequent implementation and operation of a more comprehensive, empirically validated and practically defined pedagogic framework—hence the contribution of this chapter.

Certainly, if we are ever to see pedagogic knowledge as anything akin to that of other professional disciplines, we must transcend paradigmatic allegiance and consolidate a knowledge base that is firmly grounded in research and professional practice. Anderson et al. (1998) capture this sentiment when they argue:

What is needed more than a philosophy of education is a science of education. Modern attempts at educational improvement point back to theorists (Piaget, Vygotsky, and

> Dewey) whose theories are vague by current psychological standards and lack the strong connection to empirical evidence that has become standard in the field. (p. 237)

Similarly, Stigler and Hiebert (1999) made the observation that:

> The teaching profession does not have enough knowledge about what constitutes effective teaching, and teachers don't have a means of successfully sharing such knowledge with one another. (p. 12)

Perhaps the most striking quote in this context is from Peter Drucker (1999) who argued that teaching is:

> …the only major occupation of man for which we have not yet developed tools that make an average person capable of competence and performance. In teaching we rely on the "naturals," the ones who somehow know how to teach.

However, within this context of apparent confusion about what constitutes 'effective teaching' or 'good pedagogy', there is increasing recognition of a substantive and validated research base that is beginning to constitute a 'science of learning'. Marzano (1992) argued that:

> …over the past 3 decades, we have amassed enough research and theory about learning to derive a truly research based-model of instruction. (p. 2)

More recently, Darling-Hammond and Bransford (2005), from surveying the research findings, concluded that:

> There are systematic and principled aspects of effective teaching, and there is a base of verifiable evidence of knowledge that supports that work in the sense that it is like engineering or medicine. (p. 12)

The paradox of contested paradigms on the one hand, and substantive knowledge bases on the other hand relating to teaching, provides little comfort to busy teaching professionals who seek practical guidance in this increasingly difficult endeavour we call teaching. An interesting analogy is to be found in Martin's (2009) conception of the "knowledge funnel" (Fig. 3.1: Martin's Knowledge Funnel) in which he depicts a process in which phenomena in the world can move from being a 'mystery' (experienced in some way but not understood) to a 'heuristic' (understandable in good part) and finally to 'algorithmic' (fully understood, predictable and controllable). For example, before the identification of HIV, the death of many people from AIDS was a mystery. Once identified, subsequent understanding of the aetiology and behaviour of the virus, as well as research, has led to significant improvements in controlling HIV in terms of development to full blown AIDS. However, as there is no cure, our knowledge is still far from algorithmic—so to speak. Martin's conception of heuristic is particularly important in this context for understanding effective teaching:

> Heuristics represent an incomplete yet distinctly advanced understanding of what was previously a mystery. But that understanding is unequally distributed. Some people remain stuck in the world of mystery, while others master its heuristics. The beauty of heuristics is that they guide us toward a solution by way of organized exploration of possibilities. (p. 12)

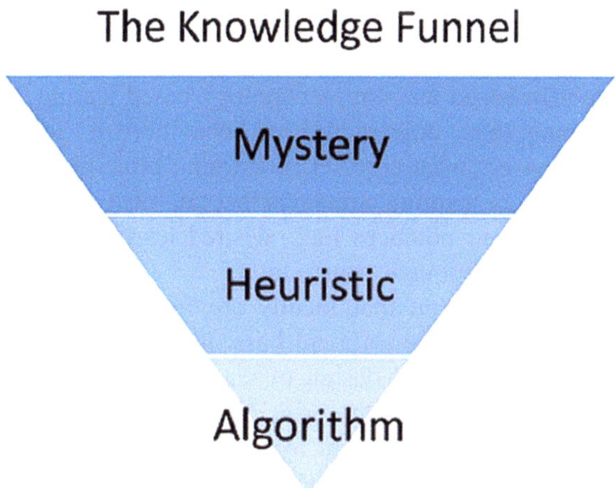

Fig. 3.1 Martin's knowledge funnel

From a science of learning position it can now be confidently argued that teaching can be seen more in terms of heuristics than a mystery. Unfortunately the paradigm debate still perpetuates a situation in which many (both within and outside the teaching profession) are still rooted in various 'genres of mystery'. The real challenge now is to critically derive these heuristics and validate and refine them through empirical study and critical reflective professional practice. This is the basis of the science of learning approach presented here. The emergent framing is depicted in the following section of this chapter.

3.3.4 The Core Principles of Learning

The core principles of learning offered in this section have been derived from an extensive review of the literature on human learning, as well as comprehensive studies on effective teaching professionals in a range of educational contexts. They are not meant to be exhaustive or summative, and they are always mediated by the situated context in which learning occurs. It will be apparent that some are explicit in CDIO standards relating to teaching, learning and assessment.

Furthermore, it is important to bear in mind that while each principle focuses attention on a key area relating to effective pedagogy, they are not discrete or separate in that they should be considered independently of each other. In fact, they are mutually supporting, interdependent and potentially highly synergetic. As Stigler and Hiebert (1999) highlight:

> Teaching is a *system*. It is not a loose mixture of individual features thrown together by the teacher. It works more like a machine, with the parts operating together and reinforcing one another, driving the vehicle forward. (p.75)

From this standpoint, the question is not so much which strategies and methods are most appropriate to facilitating desired learning (e.g., lecture, demonstration, problem-based learning, technology-based learning) or the paradigm or model of learning (e.g., constructivism, behaviourism, andragogy). Rather, the essential question is: 'what core set of learning principles can usefully inform lecturers as they design learning strategies that are most effective in the context in which they conduct their practices (e.g., desired learning outcomes, student group, environmental and resource constraints)?'

It is important that faculty are fully conversant with the principles as this constitutes the foundational base of a solid *Pedagogic Literacy*, which is fundamental to the good design of learning in any context. As faculty develop competence in applying the principles in their practices, this will support continual professional development towards becoming effective and creative designers of learning. From practice over time, given motivation and commitment, they will develop a level of *Pedagogic Competence* consistent with a truly professional model of teaching. This represents our best present understanding and application of the heuristics of effective teaching and learning. Will teaching ever be open to conceptualization at the algorithmic level? That's very unlikely as teaching has a fundamentally situated and creative thematic, but the notion of moving down - albeit with caution - a knowledge funnel that leads to more valide and reliable practices has much of merit.

The following is a summary of each of the 10 core principles. Ten was never an intention, nor a tablet of stone. Some summaries are more extensive as they are less self-explanatory and involve more conceptual understanding regarding their range of impact on specific aspects of the learning process.

3.3.4.1 Core Principle 1: Learning Goals, Objectives and Expectations are Clearly Communicated

The importance of clear goals and objectives, as well as clear explanations of what is involved in meeting them is well documented. Ramsden (1992) brings home their importance when he stated:

> It is indisputable that, from the students' perspective, clear standards and goals are a vitally important element of an effective educational experience. Lack of clarity on these points is almost always associated with negative evaluations, learning difficulties and poor performance. (p.127)

To be able to provide students with clear learning goals and criteria for success—making the learning visible—teachers themselves must firstly model the intended learning that are to be the goals for student achievement. As Hattie (2009) argues:

> Teachers need to know the learning intentions and success criteria of their lessons, know how well they are attaining these criteria for all students, and know *where to go next* in

> light of the gap between current students' knowledge and understanding and the success criteria. (p. 36–37)

Guidance on what is to be learned, the performance standards required, and how to go about this learning provides students with a clear structure from which to plan and monitor their learning. This may seem as obvious as not passing the ball across one's defensive line in football, but it's surprising how many top professional footballers still fall foul of this—not to mention local amateur park players.

3.3.4.2 Core Principle 2: Learners' Prior Knowledge is Activated and Connected to New Learning

Students come to the learning situation (whether it be the classroom or elsewhere) with preconceptions about how the world works, based on their life experiences, beliefs and values. This may involve specific understandings and misconceptions, motivational dispositions as well as levels of competence in particular related skills (e.g., reading, thinking). However, as Shulman (1991) rightly points out:

> All new knowledge gains its form and meaning through its connection with pre-existing knowledge and its influence on the organization and reorganization of prior knowledge. (p. 10)

Prior knowledge then is the lens through which students will perceive and react to new information provided in a learning event. If prior learning is inaccurate, incongruent or limited, it is likely to interfere with the meaningful integration of the new knowledge presented. Ausubel (1978) went as far as arguing that:

> If I had to reduce all of educational psychology to just one principle, I would say this: the most important single factor influencing learning is what the learner already knows. Ascertain this and teach him (sic) accordingly. (p. 163)

Making students' prior knowledge explicit helps not only to deal with misconceptions, facilitate better linking of new knowledge to existing knowledge structures, but also saves an enormous waste of time in terms of duplicated learning (Nuthall 2005) and boredom for students, as well as frustration for teachers.

3.3.4.3 Core Principle 3: Motivational and Attentional Strategies are Incorporated into Learning Designs

Motivation initiates, directs and maintains learning behaviour. According to research from cognitive neuroscience, motivation is governed by three broad principles—pleasure seeking, pain avoidance and novelty (e.g., Cloniger 1997). Students who perceive classroom learning as painful and boring are unlikely to contribute much, except to absenteeism rates and disruptive behaviour.

Similarly, getting good attention from students is essential for learning. As Sylwester (1998) points out:

> It's biologically impossible to learn anything that you're not paying attention to; the attentional mechanism drives the whole learning and memory process. (p. 6)

Csikszentmihalyi (1990) has argued:

> The shape and content of life depends on how attention has been used....Attention is the most important tool in the task of improving the quality of experience. (p. 33)

Motivation and attention are very much connected in the world of the classroom. When learners are motivated, they are much more likely to give a higher level of attention than in situations when motivation is poor. They are also more likely to put effort into the learning process—especially when difficulties are encountered.

Motivation is influenced by a wide range of interacting factors, such as personal beliefs, perceived usefulness and interest in the learning or what it will lead to. Students who believe that the learning experience may result in satisfying some aspect of personal need are more likely to participate meaningfully in the learning process.

The implication of this core principle is that the design of learning experiences must involve not only the actual subject knowledge involved, but also ways to generate and sustain learner motivation and attention. As Wlodkowski (1999) argues,

> ...if something can be learned, it can be learned in a motivating manner. (p. 24)

In today's teaching context, where we now are competing for student attention with an increasing number of potentially greater motivating distracters both inside and outside the classroom, the need for greater creativity in the ways in which we are able to create more interesting and motivating learning experiences is becoming more salient. The days of stand and deliver content from the standard text is truly another form of educational Jurassic Park. The need for more highly effective and creative teaching will no longer be a 'nice to have' for the few who can, but a necessity for the mainstream teaching force.

3.3.4.4 Core Principle 4: Content is Organized Around Key Concepts and Principles that are Fundamental to Understanding the Structure of a Subject

Understanding is about making personal meaning of knowledge and seeing how it is used in real world application and problem-solving. When students have developed a good understanding of a topic, they will have acquired accurate representation of the concepts, principles and procedures involved, which will facilitate effective and efficient retrieval and application of knowledge. Berliner's (1987) description of the benefits of good understanding in a particular field or domain is particularly informative in this context:

> Individuals possessing rich, relatively complete schemas about certain phenomena need very little personal experience to learn easily, quickly, and retain well information pertaining to those phenomena. A well-developed schemata allows very efficient learning from verbal and written discourse on a topic about which much is known. (p. 61)

Similarly, as Pugh and Bergin (2006) point out:

> …for students to access and apply their learning, they need to possess deep-level, connected knowledge structures. That is their knowledge needs to be conceptually deep, cohesive, and connected to other key ideas, relevant prior knowledge, multiple representations, and everyday experiences. (p. 148)

Teaching faculty therefore need to identify and teach the key concepts, principles and procedures that are fundamental to understanding in the topics they teach, what McTighe and Wiggins (2000) refer to as the 'big ideas' and 'essential questions'. This is increasingly important in the context of exponentially increasing knowledge in most domain fields.

Furthermore, it has long been recognized that different subject areas, by their very nature, lend themselves to different teaching and learning approaches in terms of effective student learning. For example, Shulman (1991) argues that teachers require *pedagogic content knowledge*, the ability to understand how their particular disciplines are most effectively taught. To quote Shulman in this context:

> When was the last time you saw a problem set in the study of Hamlet? Or in Asian History? Can you have guided practice in a poem? Or for evolutionary theory? I would argue that we have, reflected in the differences among the disciples, different ways of knowing that are tied to different ways of teaching. (p. 5)

3.3.4.5 Core Principle 5: Self-Directed Learning is Promoted Through Developing Good Thinking

This core principle highlights the important cognitive processes of thinking, which underpin our ability for self-regulation. The important role of thinking, its key components, working, and relationship to knowledge have been documented in Chap. 2. The summary by Jensen (1996) reinforces its importance in this context:

> The best thing we can do, from the point of view of the brain, is to teach our learners how to think. (p. 163)

Of particular significance is the role of metacognition, which acts as the regulatory component of the cognitive apparatus. Extensive research fully supports the use of a metacognitive approach to instruction (e.g., where learners are encouraged to monitor and evaluate the quality of their thinking, and use a range of learning-to-learn strategies to meet personal goals), as a means to significantly enhance learning effectiveness and efficiency (Bransford et al 1999; Hattie 2009).

Furthermore, we are also very aware of the impact of beliefs and emotions as potential inhibitors or facilitators of good thinking and learning attainment. For example, Dweck (2006) found that students who believed that their learning, even intelligence, improved through personal effort, achieved higher levels of attainment than those who have more fixed, innate, views of learning and intelligence. Similarly, Bandura (1997) observed that students who believed they were capable of meeting desired goals (self-efficacy) were much more likely to take on the required learning tasks, put in the necessary effort and achieve success, than those who lacked self-efficacy. Given the necessity of managing such 'non- cognitive' human characteristics that impact our thinking, performance and attainment, metacognition becomes the central overarching cognitive capability in self-regulation. Without effective metacognition, internal mental activity is likely to be chaotic and may indeed reflect Apter's (2001) description of human consciousness:

> ...everyday life, as it is experienced, is a tangled web of changing desires, perceptions, feelings, and emotions that filter in and out of awareness in a perceptual swirl. (p. 33)

In this context, the development of good thinking, with particular emphasis on metacognition, must be a priority in the development of human capability. Students need to understand how the mind works and what can be systematically done and how to improve the capability for better self-regulation, both for academic success and personal effectiveness in other areas of life.

We are being naive in making an assumption that effective thinking and self-regulation will naturally occur for most of our students, simply by encouraging or telling them to do so. Without sufficient foundational knowledge and skill in good thinking, as well as an understanding on how emotions, beliefs and other vagaries of the human mind influence such capability, many will lack the necessary understanding and competence to self-regulate effectively. Students who are put into situations of autonomy without preparation, may self-regulate their activities to the local canteen to play pool rather than go through the necessary thinking processes, self-management activities and appropriate use of learning strategies.

3.3.4.6 Core Principle 6: Instructional Methods and Presentation Mediums Engage the Range of Human Senses

Mental activity is stimulated through our five senses, with the visual sense being the most powerful. As the old saying goes, "a picture paints a thousand words".

Research shows that the greater the combination of our senses that are stimulated in learning, the more successful the learning is likely to be. For example, it is estimated that when we see and hear something, this doubles the sensory impact compared to just hearing it. Direct experience will increase the impact further, and, teaching it, further still. Actually assessing something has probably even greater

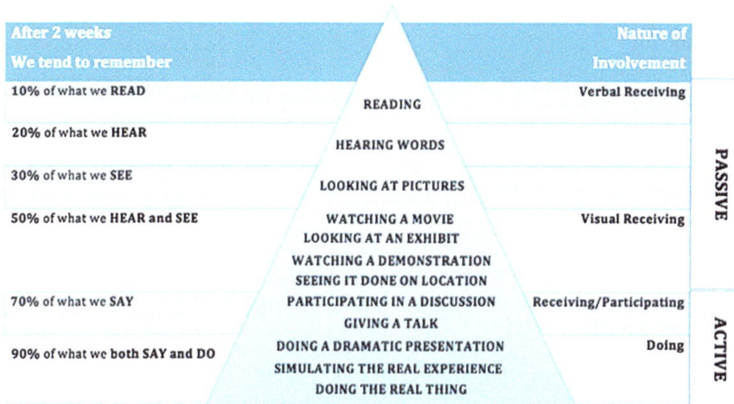

Fig. 3.2 Edgar Dales 'cone of learning'

impact than teaching it. Edgar Dales 'Cone of Learning' (Fig. 3.2 above) is a famous illustration of how different senses and activities affect the learning process. The percentages are, of course, only arbitrary.

Also, a number of researchers (e.g., Dunn and Griggs 2000; Kolb 1984) have argued that individual learners have sensory preferences (e.g., visual, auditory, kinaesthetic) in terms of how they best acquire information. Certainly, people have preferences for the ways in which they learn, though whether these can be seen to reflect different learning styles as such is contested, as noted earlier.

In today's multi-media and internet rich resource pool, finding resources to engage the range of senses becomes an easier task for the creative teacher. However, we must bear in mind that today's learners, so familiar with the internet, will not simply give attention to 'bells and whistles' multi-media. Chapter 5 explores the affordances that information communication technologies can offer for enhancing student learning, when employed from a solid pedagogic framework.

3.3.4.7 Core Principle 7: Learning Design Takes into Account the Working of Memory Systems

As knowledge is increasing almost exponentially and society is changing more rapidly, educationalists are facing an ever difficult paradox. In basic terms, there is more and more to learn within disciplines, as well as the diversity of knowledge areas and skills required for working and living in the twenty-first century. There is a call for both depth and breadth of knowledge and skills.

However, while human brains have potentially unlimited storage capacity by means of long term memory, all new learning has to firstly pass through working memory with its limited capacity of 7 plus or minus two bits. As Clark and Lyons (2005) point out:

> …it is in working memory that active mental work, including learning, takes place. Working memory is the site of conscious thought and processing. (p. 48)

Equally, understanding of the importance of long term memory is crucial for learning and the development of expertise. For example, Kircher et al. (2006) point out:

> …long term memory is now viewed as the central dominant structure of human cognition. Everything we see, hear and think about is critically dependent on and influenced by our long-term memory. (pp. 3–4)

Indeed, a major factor that differentiates experts from novices is that expert problem-solvers are able to draw on the vast knowledge bases in their long-term memory and quickly select the best approach and procedures for solving a given problem. As Kircher explains:

> We are skillful in an area because our long-term memory contains huge amounts of information concerning that area. That information permits us to quickly recognize the characteristics of a situation and indicates to us, often unconsciously, what to do and how to do it. (p. 4)

Expertise, then, enables a better and quicker understanding of a situation—perceiving what is relevant and useful for the task in hand. This enables the expert to do many things quickly and automatically, releasing time to be more situationally responsive and creative. In the context of teaching Turner-Bisset (2001) noted:

> Expert teachers are able to read and process the complex mass of information which any classroom provides, much more rapidly and meaningfully. (p. 69)

The design and delivery of instruction (in whatever form) must negotiate the working of memory systems and ensure practices work in consonance and *not* at variance with such systems. For example, information needs to be presented in manageable chunks and learners are given the necessary time to organize and make meaning of it in working memory, therefore enabling effective transfer to long term memory. Equally important is the systematic periodic review from long-term back into working memory in order to firmly establish and maintain its retention in long term memory.

3.3.4.8 Core Principle 8: Learner Competence is Promoted Through Active and Experiential Learning

The development of competence involves more than memory and understanding, but the critical synthesis of related knowledge areas, skill sets and attitudes orientated to a specific performance area. McTighe and Wiggins (2000) point out that there are fundamental tasks—"core tasks" that relate to the most important performance demands in any field. For example, a core task in science is to design and debug a controlled experiment from scratch. They go as far as to argue that:

> Core tasks with authentic challenges embody our educational aims. (p. 78)

Such activities provide an essential experiential context to learning in which students can form understandings, test out their learning in real situations, observe and reflect, further learn, etc. This needs no more illustration as it is foundational to CDIO pedagogy and has been sufficiently illustrated and contextualized earlier in the chapter.

3.3.4.9 Core Principle 9: A Psychological Climate is Created Which is Both Success Orientated and Fun

Learning is as much a social and emotional process as a cognitive one. Significant aspects of the learning environment, especially interactions with tutors and peers can play a significant part in how learners feel psychologically and in their orientation to learning. As Ornstein and Behar (1995), from research, conclude that:

> . …the most effective teachers endow their students with a "you can do it" attitude, with good feelings about themselves, which are indirectly and eventually related to cognitive achievement. (p. 86)

Bain (2004), from an extensive study of "best college teachers", noted that such teachers set and expected high standards from their students and communicated a strong trust in their ability to meet them.

Fun or humour, were certainly not significant features of my school experience; well not in classroom time. It seemed that learning was a very serious business and anything resembling a joke was akin to classroom disruption. As a Cockney from East London, I have always felt that humour was one of the most important aspects of human experience. Now, such face- validity is being supported by a wide range of research (e.g., Garner 2006; Lei et al. 2010). Far from limiting the learning experience, humour is now seen to have many positive impacts, such as:

- Refreshing the brain
- Creating mental images that retain learning
- Reinforcing desired behaviour and makes classroom management easier
- Developing positive attitudes
- Promoting creativity
- Contributing to the enjoyment of teaching.

The importance of fostering the social and emotional aspects of learning which shape the psychological climate has also been documented by Jensen (1996):

> Learners in positive, joyful environments are likely to experience better learning, memory and feelings of self-esteem. (p. 98)

John Dewey (1988) captures this core principle most definitively when he wrote:

> It is possible to be playful and serious at the same time, and it defines the ideal mental condition.

In the context of post compulsory education, especially where some faculty may see themselves primarily as subject experts and actual teaching as the more tertiary rather than *the* core professional activity, these social and emotional features of the learning experience are sometimes neglected. In reality, they are central to the learning process.

3.3.4.10 Core Principle 10: Assessment Practices are Integrated into the Learning Design to Provide Quality Feedback for Learners

This is very much in line with CDIO Standard 11: Learning Assessment. It is now clearly recognized that assessment is not simply a means to measure learning that has already occurred, but is a major facilitator in the learning process itself. As Boud (1988) illustrates:

> There have been a number of notable studies over the years which have demonstrated that assessment methods and requirements probably have a greater influence on how and what students learn than any other single factor. This influence may well be of greater significance than the impact of teaching or learning materials. (p. 35)

Well used assessment methods and processes significantly influence learning in a number of related ways. Firstly, they direct learning towards the desired learning outcomes and provide clear guides to performance criteria and standards.

Secondly, when used to support the learning process (formative assessment), it provides the essential feedback to, as Hattie (2009) notes, "…reduce discrepancies between current understandings and performance and a learning intention or goal" (p. 175).

Winnie and Butler (1994) capture the power of effective feedback in terms of:

> information with which a learner can confirm, add to, overwrite, tune, or restructure information in memory, whether that information is domain knowledge, metacognitive knowledge, beliefs about self and tasks, or cognitive tactics and strategies. (p. 5740)

In summary, there is much of merit in the learning stakes for clear, concise and timely feedback:

- clarifying what good performance is (e.g. goals, criteria, standards)
- identifying gaps in performance and specific learning needs
- closing the gap between current and desired performance
- positive beliefs and self-esteem
- the development of self-assessment in learning
- appropriate modification of instructional strategies.

3.3.5 *Using Core Principles Thoughtfully: The Fly Fishing Analogy*

For the uninitiated, fly fishing involves a fairly sophisticated fishing technique in which an artificial fly is cast to catch trout. However, whether or not the fisherperson catches trout, involves much more than this. The types of fly, the environmental conditions, species of trout, and how deep to let the fly sink and at what pace, are some of the critical considerations in catching trout. The expert fisherperson negotiates these almost intuitively and catches fish regularly. Suffice to say, as a novice fly-fisherman, I caught few trout and never reached any great heights of expertise.

Fly fishing is a useful analogy for modelling the design of effective teaching, in that both are based on solid knowledge bases relating to the design and conduct of the respective activities. Similarly, they are also mediated by the situated context in which they are enacted in that both the fly fisherperson and the teacher have to deal with the here and now situation as it is framed, select methods and resources, and create strategies to try to produce good results—whether defined in terms of 'trout caught' or 'students taught'.

It is important to understand, then, that the core principles of learning require a thoughtful application to relevant situated factors in the construction of effective teaching and learning strategies. These situated factors include such considerations as:

a. Learning Outcomes
   Different types of learning outcomes require different learning designs. For example, outcomes that require primarily the acquisition of factual content knowledge for procedural use will require a design quite different from that which seeks to promote a range of types of thinking.
b. Learner Characteristics
   Learners can differ in many ways, most noticeably in terms of motivations and competence levels. They may also have differing orientations and preferences in terms of how they learn. Certainly learners with little intrinsic motivation and limited underpinning knowledge in a specific area will pose different pedagogic challenges than highly motivated and competent learners.
c. Learning Context and Resource Availability
   A good design on paper will not work if the learning context cannot accommodate it (e.g., equipment/resources necessary are unavailable). In short, whether or not to use a strategy for meeting certain learning outcomes may be as much a resource availability issue as it is a pedagogic one.
   What this all means in practice is that the effective and creative design of learning involves a thoughtful consideration of both core principles of learning as well as the relevant situated factors involved. As Darling-Hammond and Bransford (2005) point out:

> …teachers not only need to understand basic principles of learning but must also know how to use them judiciously to meet diverse learning goals in contexts where students differ in their needs. (p. 78)

3.4 Applying the Pedagogic Framework: Effective and Creative Learning Design

In the previous section the core principles of learning were outlined and illustrated. They represent important heuristics that provide frames—or 'rules of thumb'—in the design of teaching and learning strategies. It is important to emphasize that they are *not* algorithmic—absolute and fixed procedures to apply in all teaching contexts. However, when used thoughtfully—'the fly fishing analogy'—they provide the essential design principles for creating effective and efficient student learning experiences, irrespective of the mode of delivery (e.g., face-to-face, online, blended).

In this section, examples of how the pedagogic framework translates into specific teaching and learning strategies will be outlined and illustrated. However, firstly, let's be clear about key terminology used in relation to teaching and learning. Terms like 'learning experience', 'instructional/teaching methods', 'learning activities' and 'instructional/teaching strategy' are often used, and also often misused.

3.4.1 Learning Experience

What our students derive from their interactions with us as teaching faculty is their own 'socially constructed reality', mediated by their views of the world as much as the here and now situated context (Berger and Luckman 1967). As humans our lives constitute a continuous stream of experience (even when sleeping) and, given choices, we actively seek experiences that are personally pleasurable, novel and pain reducing (as noted earlier). In the context of teaching, if students perceive our lessons as uninteresting and not useful for their perceived needs, they are unlikely to participate meaningfully in the experience. That's why many students skip classes, don't connect with the lesson content and seek more interesting activities on their notebooks.

Essentially, as faculty we are active agents in the structuring of experiences for our students, and this takes much skill and creativity for it to be optimally effective. While we can argue about what makes the experience optimally effective, for who, in what context, etc., most of us would concur that there are features of the way teachers teach that make for interest and engagement; and there are features that make for boredom and disengagement. The use of interesting stories, humour, meaningful activities (as perceived by students), presentational style (and this includes enthusiasm) and illustrative examples are all related to better student attention, engagement and learning. The converse is also true—dull presentation and monotony quickly leads students into the 'world of bla'. While we have individualized maps of the world, there appears to be an underlying syntax in the ways people structure subjective experience (Bandler and Grinder 1990), and the range of variation in that structuring (Marton 1981), enabling us to model features in the environment that are more or less likely to impact it pleasurably, painlessly or as novel.

The science of learning approach advocated in this chapter, while recognizing differences in learning preferences, argues that there is greater merit in focusing on the empirically validated similarities in human learning. The search for useful heuristics in understanding teaching and learning (e.g., core principles of learning) are a significant part of this endeavour. From my own experiences, as a trainer and facilitator in many countries, it is readily apparent that effective pedagogical practices transcend cultural contexts. For example, Sale and Mukerji (2006) wrote:

> In our experiences of co-facilitation over several years, we were initially surprised but ultimately delighted to find that there appears to be a number of generic principles and practices that facilitate rapport and effective learning irrespective of cultural and ethnic contexts. (p. 1)

3.4.2 Instructional/Teaching Strategy/Learning Design

Instructional/teaching strategy or learning design refers to the overall plan for creating the student learning experience to promote the desired outcomes for a learning event. It typically involves a combination of **instructional methods**, **learning activities** and **resources**.

Instructional methods are, in the broadest sense, teacher directed planned structures for creating a particular type of learning experience, directed towards meeting certain learning outcomes. The traditional *core* methods used by teachers are explanation (often referred to as 'chalk and talk' or the lecture method) and demonstration. Other *supporting* methods include questioning, group work, discussion, role-plays, debates. The main feature of any method is that it provides the 'how to' for achieving the acquisition of certain knowledge or skills.

Learning activities are specific performance-based tasks used to engage students actively in the learning process, again with the intention of contributing to the attainment of desired learning outcomes. Activities are usually used in unison with methods, and can sometimes refer to the same thing. For example, cases are considered a method of instruction, but the actual *case* is an activity in itself.

Resources are essentially anything else beyond specific methods and activities that are used to support the learning process. At a more macro level, this will involve all the facilities and equipment available for supporting learning, e.g., classrooms, laboratories, online facilities as well as everyday teaching and learning aids such as presentation slides, multimedia and hand-outs. In the CDIO context, the development and use of customized learning environments—CDIO Standard 6: Engineering Workspaces—is an important part of the overall design of the learning experience. For example, we are all aware that having students sit in rows in the traditional classroom setting is not conducive for active learning and student discussion, etc. Being able to create learning spaces that model real world work contexts, simulating the key activities done by professionals in their everyday work and problem solving, is central to providing a conducive learning environment that facilitates active and experiential learning. The basic idea is that the designed learning spaces will not only provide the necessary physical resources used in real work contexts, but also foster the kinds of interactions necessary for developing the range of skills documented in Chap. 2. In SP a number of innovative learning spaces have been developed across all the participating schools. One that is particularly innovative is the War Zone (Fig. 3.3) in which internet security protection skills are developed through a learning space that simulates cyber-attack and defence.

The notion of *strategy* applies both across a large unit of study (e.g., a module) or a smaller unit such as a lesson. Invariably, the range of methods, activities and resources that can be strategically employed is significantly greater across longer learning events, but it should always be a key feature in planning the learning experience for students.

The design of effective instructional strategies requires a sound *pedagogic literacy* which has been the major focus of this chapter. I am often intrigued to hear experienced teachers pose question such as:

- Are lectures a poor way to teach?
- Is cooperative learning effective?
- Will a blog help my students to learn?

The very asking of such questions often reveals a lack of pedagogic literacy. They can only be usefully answered once subjected to sound pedagogic thinking, which might go something like this in relation to whether or not to use a lecture:

- What learning outcomes do I want my students to meet?
- Are these best achieved through what I can do in a lecture?

Fig. 3.3 War Zone Learning Space at School of Digital Media & Infocomm Technology

- How might this particular group of students respond to a lecture in terms of interest and desired learning outcomes?
- What are alternative methods, activities and resources—are they likely to get better results, on what basis, etc.?

If after such deliberations, a lecture appears to be an effective method in this context, it's then a question of how best to design it for the student group involved. In other words, simply choosing an appropriate method is far from job done. By applying the core principles of learning, it should be apparent that many are very applicable to making a lecture a more engaging and meaningful experience for students (e.g., ' incorporate motivational and attentional strategies', 'connect to prior knowledge' 'engage a range of senses', 'organize content around key concept and principles', 'create a positive psychological climate'). Failure to leverage on some of these in the lecture design and delivery will mitigate its effectiveness. However, when a lecture meets many of the core principles outlined, and the method is not over used, it is a powerful way to present key new information, provide structure to large bodies of content, separate the 'wood from the trees', simplify and motivate. In this situation the 'sage on the stage' is far better than the 'guide on the side', as the analogy by Bransford et al. (1999) quite aptly illustrates:

> Asking which teaching method/technique is best is analogous to asking what tool is best—a hammer, a screwdriver, a knife, or pliers. In teaching, as in carpentry, the selection of tools depends on the task at hand and the materials one is working with. (p. 22)

In summary, then, there are a lot of planning considerations to bear in mind when designing and creating the actual learning experience for students. The core principles of learning provide the means to conduct this processes in effective, efficient and creative ways. They are essential for the design of the overall strategy (at whatever level), which involves the selection of methods, activities, and resources, and how they are integrated and used in the situated context. How well this process is conducted determines in large part the impact of the experience in terms of student attention, engagement and learning.

3.4.2.1 The Design of Learning Activities

A key aspect of the CDIO pedagogic approach involves the use of real world learning activities or performance tasks, often in the form of integrated capstone and design-build projects (Standard 5: Design Implement Experiences and Standard 7: Integrated Learning Experiences). Performance-based learning tasks have become increasingly popular across a whole range of educational sectors as they can provide:

- authentic learning opportunities that mirror real world activities and make learning more meaningful
- integrated learning activities that naturally combines subject knowledge, types of thinking and other process skills

- a more authentic assessment of actual competencies than traditional pencil and paper tests
- a means of developing intrinsic motivation by offering students more autonomy and control in their learning
- a framework for both a collaborative and active learning approach, whereby students have to solve problems through information resourcing, good thinking and practical application.

However, the use of such activities, whether in the form of projects, cases or simulations, need to be designed, implemented and managed from a sound pedagogic base, as described in this chapter. Furthermore, as with any experience in life, too much of one thing tends to create habituation and eventual boredom. While meaningful and interesting project activities can result in enhanced student learning, as identified above, too many projects over time may soon lead students to perceive these activities as just another part of the 'daily grind'.

A major skill area of CDIO is Personal Skills and Attitudes, which subsume skill sets relating to good thinking and managing learning. It is important therefore, to ensure that specific types of thinking (e.g., analysis, comparison and contrast, inference and interpretation, generating possibilities, and metacognition) are clearly and appropriately infused into stages of performance-based learning activities. As emphasized in Chap. 2, students need to clearly understand what good thinking actually entails (the cognitive heuristics involved), have opportunities for active and experiential application in real world contexts, as well as receive clear and useful feedback from expert professionals. Sheppard et al. (2009) are correct when they argue that:

> …teachers have to make their own intellectual processes (their performances) visible. This means that the teacher-expert has to make visible to learners the otherwise invisible processes of thinking that underlie complex cognitive operations at the heart of engineering thinking. Teachers have to articulate and demonstrate rather than assume the thought processes they want students to learn.
>
> …Then student's efforts to replicate these thought processes need to be made visible so that the teacher can see where the learner is on and off track, in order to provide appropriate coaching and feedback. (p. 188)

Similarly, Ritchhart et al. (2011) point out:

> Making students' thinking visible serves a broader educational goal as well. When we demystify the thinking and learning process, we provide models for students of what it means to engage with ideas, to think, and to learn. In doing so, we dispel the myth that learning is just a matter of committing the information in the textbook to one's memory. (p. 28)

In the SP context, the following broad design model has been used across schools in the design of real-world performance-based learning activities that seek to foster the integration of thinking and other process skills with technical subject content:

Step 1: Identify the key technical content areas, types of thinking and other process skills to be incorporated in the learning activity

For this step it is important to:

- select specific topic areas in the curriculum that contain key knowledge essential for building understanding of the subject. For example, central concepts, principles and procedures
- identify the types of thinking that are important for promoting student understanding and subsequent competence in these topic areas (analysis, comparison and contrast, inference and interpretation, evaluation, generating possibilities, metacognition)
- identify other process skills that are important for promoting learning in the identified areas (communication, teamwork, learning, etc.)

An example of framing the main knowledge and skill components for an electrical and electronic engineering project is presented in Table 3.1.

Step 2: Produce the learning task activities

This stage involves constructing the project task itself. It is essential to ensure that:

- the task incorporates the application of knowledge, skills and processes specified in Step 1.
- the activities mirror, as far as possible, real world applications
- it is sufficiently challenging, but realistically achievable in terms of student's prior competence, access to resources, and time frames allocated
- successful completion involves more than one *correct answer* or more than one *correct way* of achieving the correct answer
- clear notes of guidance are provided which:
  - identify the products of the project task and what formats of presentation are acceptable (e.g. written report, oral presentation, portfolio)
  - specify the parameters and scope of the activity (e.g., time, length, areas to incorporate, individual/collaborative, choice is permitted, resource access, support provided, etc.)

Table 3.1 Main knowledge and skills components for an electronic engineering project

Project components		
Subject knowledge	Types of thinking	Other process skills
– Circuit design and integration principles – Circuit building – Use of sensors	– Generating possibilities relating to circuit design – Analysis—part-whole relationships of sensors in an integrated circuit – Compare and contrast—previous options and new options generated – Making inferences and interpretations from data relating to the behavior of sensors in an integrated circuit – Evaluation of interesting options in relation to derived criteria	– Oral and written communication – Teamwork

- cue the types of thinking and other desired process skills
- spell out key aspects of the assessment process and criteria.

Box 3.1 is an example of a project task used in a mechanical engineering module based on the learning outcomes example identified in Step 1.

Box 3.1

Project #2—Moving Car Transit

Notes of Guidance

Objective:

This project requires you to design, build and test electronic circuits necessary to control a range of movements in a model car (e.g., mode forward, reverse and stop).

The project is to be done in groups of 3–4 and will be completed by xxxx.

Scope:

To meet the project requirements you will need to:

- Form a work team and organize the necessary activities you will need to do in order to complete the technical requirement specified below. (Note: it is important that your team identifies clear roles and responsibilities, distributing and coordinating various tasks appropriately, and is able to operate as a high performing team).
- Build and test the following circuits:

  1. Light Dependent Sensor Circuit—to detect the station.
  2. Counter and Display Circuit—to display the Station number on the 7—segment LED display.
  3. Motion Control Circuit—to activate the motor and move the car in forward or reverse direction.
  4. Voltage Regulator Circuit—to provide 5V dc supply.

- Design a Counter Limiting Circuit that is able to integrate the above circuits, enabling the car to move forward to any Station, reverse automatically and stop after hitting an emergency switch (micro switch) in both forward and reverse directions.
- Incorporate additional specific performance and/or aesthetic features which may differentiate your car from the rest (e.g., can do extra movements, perform faster in certain movements, has novel/attractive appearance).

This ICA project comprises 50 % of the marks weighting for this module.

The distribution of marks for the various project components is contained in Table 1

Table 1

Assessment components	Mark weighting in percentage (%)
Plan, Build and Test Circuits 1–4	40
Counter Limiting Circuit Design	20
Creativity (e.g., enhanced functionality, aesthetics)	20
Teamwork (e.g., goal setting, management of team-roles and responsibilities, dealing with conflict/challenges)	10
Communication (e.g., clarity and cohesiveness of explanation)	10
Totals	100

In developing the project task brief and preparing the necessary logistic requirements, faculty must decide if the problem and the level of constraints are appropriate for the particular student cohort. In making this decision, it is important to refer to Step 1, where the desired skills and knowledge have been identified. More structured projects with clearly identified constraints are most useful initially where students lack experience in doing real world projects. Furthermore, another advantage of such projects is that they are easier to align more precisely to module syllabus learning outcomes, hence making the assessment process more reliable and resource effective. Finally, in terms of administrative demands, this type of project is likely to be less problematic as all necessary logistics and preparation can be made ready before the project commences.

However, the limitations of a more structured project brief is that it reduces opportunities for students to find and frame problems which are fundamental to the overall CDIO approach, especially conceiving and designing engineering systems. Also, such projects may reduce student's motivation as school assigned projects may not be perceived as meaningful and interesting. Invariably, a different set of challenges arise if students have little constraints in their choices of projects. Students may decide to work on projects of interest that are out of the knowledge domain and not related to the course and module objectives. Also, the scope of projects may be too wide that they have difficulty completing significant and assessable outcomes.

Essentially, decisions relating to project types, structure and complexity must be thoughtfully negotiated by faculty over the duration of the course programme. In broad terms, the movement from high structure and constraints to more open ended ill-defined problems represents the general norm, but must always be contextualized to the learning outcomes, the student groups and resource availability.

From our experience in SP, one recommended approach is for faculty to supervise projects that guide students in finding problems. In this situation, the constraints could be broadly framed using project themes or they may be bounded by certain knowledge domains taught in the class. Faculty setting the projects must carefully negotiate a project brief that utilizes student knowledge and competences and connects to problems that are realistic for their present learning context. For example, in a Design and Innovation Project, students have been required to develop a technical artifact to harvest renewable energy using the knowledge acquired in the classroom. For this purpose, students chose to build a sun tracking solar panel that is used to collect solar energy. Guided problem finding satisfies the need for students to find personally relevant problems and at the same time allows them to work within the specified knowledge domains of module syllabi. Invariably, in this situation, the challenge becomes one of aligning the project deliverables with the module(s) designed learning outcomes.

Additionally, faculty will need to manage the additional logistic requirements as such projects typically require more hardware resources and flexibility. For example, this requires an agile purchasing system as a project may be severely compromised if the necessary parts do not get delivered on time, especially when the project only runs for a few weeks. A possible way to minimize this logistic inconvenience is to ensure that the project design has a fixed major component that may be purchased prior to the start of project and a smaller set of flexible parts that need to be purchased when the project specifications are ready.

Another facet to designing project assignments is the level of complexity of problems. Should the students work on existent, emergent or potential problems? Again, this question may only be answered in relation to the learning objectives for a specific curriculum. Existent problems are evident and there is little need to engage in problem finding. Students only need to redefine the problems in some instances, though may still display good thinking skills in coming up with creative and innovative solutions to them.

In contrast, emergent problems must be discovered before they can be solved. Students need to find and probe data for a hidden, unclear or, what in the language of design thinking is referred to as, 'latent needs'. As Brown (2009) points out, in discussing why people are often not able to make explicit what they would like from an experience:

> The basic problem is that people are so ingenious at adapting to inconvenient situations that they are not even aware of what they are doing. (p.40)

SP is presently implementing a 'design thinking' approach (e.g., Brown 2009; Rowe 1991), particularly in relation to the 'conceive' and 'design' aspects of third year integrated and multidisciplinary projects. A design thinking approach attempts to apply good thinking-conceptualized in terms of marrying good analytical thinking with more intuitive creative thinking-to meet human needs in ways that are technically feasible and viable in terms of sustainable business. Indeed, design is fundamental to the practice of engineering, as Dym et al. (2005) point out:

> Design is what engineers do, and the intellectual and thoughtful design of the engineering curriculum should be the community's first allegiance. (p. 114)

Furthermore, as Dym et al. argue, "Design is both a mechanism for learning and in itself a learning process" (p. 112). The systematic infusion of the skills and practices of design thinking enhance the 'science of learning' approach to pedagogy that has been developed in this chapter. Design involves, as Beetham and Sharpe (2007) illustrate:

> ...a systematic approach with rules based on evidence, and a set of contextualized practices that are constantly adapting to circumstances. It is a skilful, creative activity that can be improved on with reflection and scholarship. (p. 6)

The incorporation of design thinking into the CDIO curriculum approach provides a structure in which students can learn a set of core skills and practices that build a foundation for more innovative design. For example, apart from recognizing the strategic use of both critical and creative thinking skills, design thinking moves away from quantitative approaches to framing human experience in more qualitative understandings of what people actually do in the real world contexts of their personal experience. As a result greater emphasis is placed on more ethnographic or participant observation approaches to understanding people's expressed needs and, perhaps more significantly, their latent needs in relation to potential products and services. In terms of pedagogic approach, students will spend more time thinking about what data to collect, where and how, verifying the date in more qualitative experiential terms, before moving to more traditional modes of quantitative data analysis and subsequent framing of problems.

Since 2009, a multi-disciplinary project programme has been piloted in which students from various schools work on a range of multi-disciplinary projects. The programme aims to:

- expose students to design thinking and related methodologies
- encourage multi-disciplinary collaboration in working on real-world projects
- develop a humanistic dimension in user research
- enable the prototyping of ideas and concepts into potentially innovative solutions.

At present, this learning experience is not offered to all students. As it is a pilot project, a key emphasis has been on ascertaining the most viable arrangements, both pedagogically and in terms of resource arrangements.

However, a number of challenges have surfaced in terms of actual implementation, both in relation to pedagogic issues and organizational arrangements. For example, what level and type of expertise is needed for an academic faculty to effectively facilitate such projects? Is it sufficient to have one project facilitator or is it necessary to have a 'pool' of various subject specialists available? Furthermore, how are the most 'desirable conditions' viably and efficiently arranged? Similarly, such projects provide challenges for timetabling and a wide range of other resource provision arrangements. For example, in some projects, much time has been spent on achieving a deep understanding and ideation, leaving insufficient time for prototyping and 'proof of concept'. Also, as some projects tended to become focused predominantly in one domain field more than others, some students felt uncomfortable as they lacked the knowledge to fully participate.

The present plans are to leverage on an already established project management platform, which is used for Final Year Projects on most SP courses. This will provide much of the resource structuring arrangements and facilities. The challenge will be to ensure sufficient quality project concepts for students to develop the range of skills that multi-disciplinary projects could potentially yield.

3.5 Summary

This chapter has outlined the pedagogic approach of CDIO and the implication for teaching and learning practices. While the essential approach documented by Crawley et al. (2007) established the active and experiential learning base for CDIO pedagogy, it has hopefully extended the pedagogy in more substantive and specific ways. A 'science of learning' approach has been advocated, which transcends previous paradigmatic alliances to constructivism, and aligns CDIO pedagogy to the thoughtful application of an increasing body of validated knowledge relating to human learning. The core principles outlined in the chapter are illustrative of how we can design student learning experiences from a sounder pedagogic base, enhancing the potential for both greater effectiveness and efficiency in desired learning. Furthermore, student induction into the various skills and practices of design thinking provides an active and experiential context to integrate a wide range of CDIO skills in solving real world problems. The intended heightened pedagogic competence in staff capability for creative learning design is consistent with teaching as 'professional activity', and meriting similar professional development to field knowledge upgrading.

References

Anderson JR et al (1998) Radical constructivism and cognitive psychology. In Ravitch D (ed.) Brookings papers on education policy 1998, pp. 227–255. D.C. Brookings Institution, Washington

Apter MJ (ed) (2001) Motivational styles in everyday life: a guide to reversal theory. American Psychological Association, Washington

Ausubel DP et al (1978) Educational psychology: a cognitive view. Holt, Rhinehart and Winston, New York

Bain K (2004) What the best college teachers do. Harvard University Press, Cambridge

Bandler R, Grinder J (1990) Frogs into Princes: the introduction to neuro-linguistic programming. Eden Grove Editions, Middlesex

Bandura A (1997) Self efficacy: the exercise of control. Worth Publishers, New York

Berliner DC (1987) Ways of thinking about students and classrooms by more and less experienced teachers. In: Calderhead J (ed) Exploring teachers' thinking. Cassell, London

Beetham H, Sharpe R (2007) Rethinking pedagogy for a digital age: designing and delivering e-learning. Routledge, London

Bransford J et al (1999) Brain, mind, experience & school. National Academy Press, Washington

Berger PL, Luckmann T (1967) The social construction of reality: a treatise in the sociology of knowledge. Anchor Books, New York

Bonwell C, Eison J (1991) Active learning: creating excitement in the classroom. ASHE-ERIC Higher Education Report No. 1

Brown T (2009) Change by design. HarperCollins, New York

Boud D (ed) (1988) Developing student autonomy in learning. Kogan Page, London

Chickering AW, Gamson ZF (1987) Seven principles for good practice in undergraduate education. American Association of Higher Education Bulletin, pp 3–7

Clark RC, Lyons C (2005) Graphics for learning: proven guidelines for planning, designing, and evaluating visuals in training materials. Pfeiffer, San Francisco

Cloniger CR (1997) A psychobiological model of personality and psychopathology. Jpn J Psychosom Med 37:91–102

Coffield F, Mosley D, Hall E, Eccleston K (2004) Learning styles and pedagogy in post-16 learning: a systematic and critical review. Learning and Skills Development Agency

Crawley E, Malmqvist J, Ostlund S, Brodeur D (2007) Rethinking engineering education. Springer, New York

Csikszentmihalyi M (1990) Flow: the psychology of optimal experience. Harper Row, New York

Darling-Hammond L, Bransford J (2005) Preparing teachers for a changing World: what teachers should learn and be able to do. Jossey-Bass, San-Francisco

Dewey J (1998) Quote. http://libertygroup.tripod.com/quotes.html. Accessed 12 Nov 2012

Drucker P (1999) Quote. (http://thinkexist.com/quotes/peter_f._drucker/2.html. Accessed 18 Aug 2010

Dunn R, Griggs SA eds (2000) Practical Approaches to using Learning Styles in Higher Education. Bergin and Garvey, Connecticut

Dweck CS (2006) Mindset: the new psychology of success. Ballantine, New York

Dym CL et al (2005) Engineering design thinking, teaching, and learning. J Eng Educ 94(1):103–120

Edstrom D et al (2007) Teaching and learning. In: Crawley EF (ed) Rethinking engineering education: the CDIO approach. Springer, New York

Garner LA (2006) Humor in pedagogy: how haha can lead to aha!. Coll Teach 54:177–179 (Accessed 25 Sept 2011)

Hattie J (2009) Visible learning. Routledge, New York

Izumi TL, Evers WM (2002) Teacher quality. Hoover Institutional Press, San Francisco

Jensen E (1996) Brain based learning. Turning Point Publishing, Del Mar

Johnson DW, Johnson RT, Johnson-Holubec E (1994) Cooperative learning in the classroom. Association for Supervision & Curriculum Development
Kircher PA et al (2006) Why minimal guidance during instruction does not work: an analysis of the failure of constructivist, discovery, problem-based, experimental, and inquiry-based teaching. Educ. Psychol. 41(2):75–86
Kimble GA (1961) Conditioning and learning. Prentice-Hall, Englewood-Cliffs
Knowles M (1984) Andragogy in action. Jossey-Bass, San Francisco
Kolb DA (1984) Experiential learning. Prentice Hall, Englewood Cliffs
Kolb DA (1995) The process of experiential learning. In: Thorpe et al. Culture and processes of adult learning. The Open University, Milton Keynes
Lave J, Wenger E (1993) Situated learning: legitimate peripheral engagement. Cambridge University Press, Cambridge
Lei SA, Cohen JL, Russler KM (2010) Humor on learning in the college classroom: evaluating benefits and drawbacks from instructor's perspectives. J Instr Psychol (Accessed 25 Sept 2011)
Littlejohn A, Pegler C (2007) Preparing for blended e-learning. Routledge, London
Martin R (2009) The design of business. Harvard Business Press, Massachusetts
Marton F (1981) Phenomenography – describing conceptions of the world around us. Instr Sci 10:177–200 (Elsevier Scientific Publishing Company, Amsterdam)
Marzano RJ (1992) A different kind of classroom. ASCD, Alexandria
McTighe J, Wiggins G (2000) Understanding by design. Association for Supervision and Curriculum Development
Mortimore P (ed) (1999) Understanding pedagogy & its impact on learning. Paul Chapman, London
Nuthall GA (2005) The cultural myths and realities of classroom teaching and learning: a personal journey. Teach Coll Rec 107(5):895–934
Ornstein AC (1995) Teaching: theory into practice. Allyn & Bacon, Needham Heights
Ornstein AC, Behar LS (1995) Contemporary issues in curriculum. Allyn & Bacon, Needham Heights
Paul RW (1993) Critical thinking. Foundation for Critical Thinking, Santa Rosa
Pratt DD (2002) Good teaching: one size fits all? (pp. 5–15). In: Ross-Gordon JM (ed) Contemporary viewpoints on teaching adults effectively, No. 93 Spring 2002. Jossey-Bass, San Francisco
Pugh KJ, Bergin DA (2006) Motivational influences on transfer. Educ Psychol 41(3):147–160
Ramsden P (1992) Learning to teach in higher education. Routledge, London
Rivers CJ, Sanders WL (2002) Teacher quality and equity in educational opportunity: findings and policy implications. In: Izumi TL, Evers WM (eds) Teacher quality. Hoover Institution Press, San Francisco
Ritchhart R, Church M, Morrison K (2011) Making thinking visible. Jossey-Bass, San Francisco
Rowe PG (1991) Design thinking. MIT Press, Massachusetts
Sale D, Mukerji S (2006) Establishing rapport in facilitating workshops in different cultural and ethnic contexts: key principles of effective human conduct and engagement. Paper presented at ERAS conference, Singapore
Sallis E, Hingley P (1991) College quality assurance systems in *Mendip Paper D20*. Coombe Lodge, Bristol
Schank R (1997) Virtual learning. McGraw-Hill, London
Sheppard SD et al (2009) Educating engineers: designing for the future of the field. Jossey-Bass, San Francisco
Shulman L (1991) Pedagogic ways of knowing. Institute of Education, London
Silberman M (2005) 101 ways to make training active. Wiley, New Jersey
Stenhouse L (1989) An introduction to curriculum research and development. Heinmann Educational Books, Oxford
Stigler JW, Hiebert J (1999) The teaching gap. Free Press, New York

Stone JE (2000) Teacher training and pedagogical methods. Hoover Institution/Pacific Research Institute Teacher Quality Conference, May 12

Sylwester R (ed) (1998) Student brains school issues. Skylight Training and Publishing Inc, Arlington Heights

Tuckman B (1995) The competent teacher (pp 57–72). In Ornstein AC (1995) Teaching: theory into practice. Allyn & Bacon, Needham Heights

Turner-Bisset R (2001) Expert teaching: knowledge and pedagogy to head the profession. David Fulton, London

Winnie PH, Butler DL (1994) Student cognition in learning from teaching. In Husen T, Postlethwaite T (eds) International encyclopaedia of education, 2nd edn, pp 5738–5745. Pergamon, Oxford

Wlodkowski RJ (1999) Enhancing adult motivation to learn: a comprehensive guide for teaching all adults. Jossey-Bass, San Francisco

Chapter 4
Assessing Learning

4.1 Introduction

> From our student's point of view, assessment always defines the actual curriculum…Assessment sends messages about the standard and amount of work required, and what aspects of the syllabus are most important.
>
> (Ramsden 1992, pp. 187–188)

One of the most significant developments in education over recent years has been an increasing focus on the importance of assessment, and the need to ensure quality in assessment practice. Two main factors have contributed to this development.

Firstly, greater pressure on educational institutions to be accountable for their products has made assessment high profile. We need to be able to justify public expenditure in terms of value for money outputs. The quality of teaching and the cost-effective use of resources are rightly important issues in this context. However, it is assessment that largely defines the value of accredited educational programmes. If assessment practices are lacking in quality, what value can be placed on the qualifications accredited? Furthermore, as Bloxham and Boyd (2007) highlight:

> Assessment is now expected to assess subject knowledge and a wide range of intellectual, professional and generic skills in a quality assurance climate that stresses reliability with robust marking and moderation methods. (p. 4)

Secondly, and most significantly for the purpose of this chapter, there is an increasing recognition of the important role that assessment plays in the learning process (e.g., Boud 1995 ; Ramsden 1992). Assessment is not simply a means to measure learning that has already occurred, it is a major facilitator in the learning process itself. As Boud (1988) illustrates:

> There have been a number of notable studies over the years which have demonstrated that assessment methods and requirements probably have a greater influence on how and what students learn than any other single factor. This influence may well be of greater significance than the impact of teaching or learning materials. (p. 35)

D. Sale, *The Challenge of Reframing Engineering Education*,
DOI: 10.1007/978-981-4560-29-0_4,

Furthermore, much research supports the view that students choose their approaches to learning rather than these being the result of innate characteristics or dispositions. For example, Prosser and Trigwell (1998) argue:

> ...approaches to learning are not stable characteristics of students. Student's *approaches* to learning do change with changes in perception of their learning situation and their perception of it can be changed by...teachers. (p. 83)

This is particularly significant if we accept the view that some approaches to learning are qualitatively better than others. Again to quote Prosser and Trigwell (1998) in this context:

> ...there are better and worse ways for students to approach their learning—a deep approach being better than a surface approach. (p. 7)

What this means in practice is that how we design and conduct our assessment must incorporate a range of quality assurance issues both in terms of quality of the learning developed as well as the validity and reliability of the assessments made. This chapter documents and illustrates our approach to providing quality of assessment practice within the CDIO Framework.

4.2 The CDIO Approach to Assessment

The CDIO approach is documented in Standard 11: Learning Assessment: Assessment of student learning in personal and interpersonal skills; and product, process, and system building skills, as well as disciplinary knowledge.

Most significantly, assessment is seen as an integral and integrated part of the overall instructional and learning process. There is a strong emphasis on formative assessment in which learning can be collaboratively monitored and evaluated by both faculty and students. The effective employment of a systematic formative assessment approach-which provides quick feedback on student learning performance, identifies learning gaps and problems, and provides essential feedback to faculty on the effectiveness of the teaching and learning strategies used-will strongly support the learning process. Such an approach is captured by Knight (1995):

> The key to the use of assessment as an engine for learning is to allow the formative function to be pre-eminent. This is achieved by ensuring that each assignment contains plenty of opportunities for learners to receive detailed, positive and timely feedback, with lots of advice on how to improve. (p. 81)

From this perspective, teaching and assessment are simply two sides of the same coin or as Perkins (1992) suggests, “Teaching, learning, and assessment merge into one seamless enterprise” (p. 176). Indeed, through collaborative formative assessment we can get to know our students and their learning preferences. We can then forge the essential communication links that foster a supportive learning relationship and build trust.

However, assessment is not an exact science and the assessment of complex real world performance, which involves the integration of a range of knowledge and skill bases, provides real challenges in terms of achieving a sufficient level of validity and reliability of assessment in a realistic efficiency context. As Gray (2007) points out:

> Finding or creating reliable, valid and appropriate assessment methods and tools matched to all learning outcomes remains a challenge. (p. 165)

Furthermore, assessment is a time consuming process, and it is unlikely that additional resources will be available in the present context of cost consciousness and already heavy workloads. As Boud (2000) warns:

> One of the traps in arguing for a shift in assessment practice is to propose an unrealistic ideal that can never be attained. (p. 159)

The reality, therefore, will be one of working smarter with existing resources. To achieve this, faculty will need to have a clear frame on what constitutes good assessment, the assessment formats, methods and strategies available, as well as the compromises that may need to be made and their impact on assessment quality.

The following sections consider three essential questions which are central to achieving quality of assessment practice:

1. What is quality assessment?
2. What specific strategies can be used to enhance the effectiveness and efficiency of assessment practices?
3. How can valid and practical assessment instruments be developed?

4.2.1 What is Quality Assessment?

Firstly, it is important to recognize that there are many different purposes for assessing student learning, and these reflect different stakeholder interests. For example, as Rowntree (1987) points out, assessment may serve any of the following purposes:

- Selection and grading
- Maintaining standards
- Diagnosing learning difficulties
- Proving feedback to support the learning process
- Evaluating the effectiveness of the teaching/learning strategy.

Furthermore, these purposes are not necessarily complimentary, and may conflict in practice. For example, while grades and standards may be of prime interest to employers, grading may do little to help students learn more effectively in the qualitative sense. In the context of CDIO, it is important that the assessment approach effectively and efficiently addresses these stakeholder interests in a balanced manner. CDIO course assessment must enable validity of the summative outcome standards as well as support the learning process through formative assessment.

4.2.1.1 Principles of Good Assessment

In terms of quality of the assessment approach, there are certain generic principles of good assessment that need to be appropriately applied in the design and conduct of assessment practices, these are:

- Validity
- Reliability
- Fairness
- Flexibility
- Sufficiency
- Authenticity.

The principles of good assessment are well documented in the literature (e.g., Haladyna 1997; Osterlind 1989; Rowntree 1987), hence only a summary reference will be made here.

Validity
This refers to a test's capability in measuring accurately what it is we intend to measure, whether this be recall of factual knowledge, understanding of concepts, competence in performance, or combinations of these learning outcomes. A major consideration in determining the validity of a test is the extent to which the evidence generated by the assessment items supports an accurate interpretation of the test score in relation to desired learning outcomes. This is typically unproblematic in the assessment of factual knowledge, but becomes more challenging in the case of assessing more complex performances involving the integration of a range of knowledge bases, skills and attitudinal components.

Reliability
This refers to the capability of a test to produce the same scores with different examiners, resulting in consistent and stable scoring of students over time. Fixed response items are typically reliable as the answers are factual and scoring is a relatively simple process. However, in more complex assessments, reliability becomes problematic, as Banta et al. (2009) note:

> As faculty members increasingly rely on applying rubrics to student work, interrater reliability becomes another matter to address. Although multiple raters may use the same

> rubric to asses student work, assessment leaders should carefully determine how consistently individual assessors are judging student work. (p. 23)

Fairness
Fairness relates to a number of considerations in assessment. However, they are all concerned with ensuring that learners, when being assessed, are provided with appropriate access to the assessment activities and are not unfairly discriminated against in the assessment process. Unfair discrimination typically means discrimination based on criteria unrelated to the assessment activity itself, for example, gender or racial characteristics. Fairness is a general concern throughout assessment, relating as much to providing learners with sufficient knowledge and time for assessment, to non-discriminatory processes in marking their work.

Flexibility
Flexibility is concerned with the process of assessment, not the standard of the assessment. Learners can display their learning in a range of ways (e.g., orally, written, demonstration, etc.), provided the evidence is validly demonstrated. Flexibility typically becomes a consideration for learners with special needs (e.g., visual/auditory impairment, second language, etc.) or untypical situations (e.g., sickness on exam day, etc.). The arrangements for flexibility are usually specified by exam boards.

Sufficiency
Sufficiency is one of the most challenging of the principles of good assessment. In most basic terms, it refers to the question of how much assessment evidence is needed in order to feel confident that a student is competent in an assessment context. This can refer to both the range of evidence and the extent to which the evidence needs to be generated. For example, how many times would a student need to demonstrate mastery of a complex and critical procedure, and in what range of contexts, before we would deem him/her competent? There are no absolute answers here, but good professional judgement and collaboration with other professionals in the field is essential for establishing realistic benchmarks.

Authenticity
Quite simply this refers to how sure we are that the work produced has been done by the student. In an examination, we can be more confident of authenticity. However, with assignments done by students in their own time, authenticity becomes a concern.

4.2.1.2 Assessment Standards

For a practical quality assurance framework to the assessment approach, *assessment standards* offer a useful guide for developing, monitoring and evaluating assessment practices. It is to be noted that while such standards are generic to the overall processes, methods and procedures of assessment practices, there is always a need to contextualize them to the particular assessment context. The following exemplar standards developed in the SP context relate to the 3 main interrelated stages of the typical assessment process:

(a) Producing and reviewing an assessment plan
(b) Judging evidence and making assessment decisions
(c) Providing feedback on assessment decisions

(a) Producing and reviewing an assessment plan

Essentially an assessment plan identifies the why, what, when, where and how of the assessment process for a module or unit of study. When well-constructed it provides a concise guide of the assessment process and components to both assessing faculty and students involved. In congruence with the CDIO aligned curriculum framework (Biggs 1996), it should result in assessment methods, instruments and procedures that both effectively develop student learning (formative assessment) and measure (summative assessment) the defined learning outcomes for the specified programme. The following criteria, which incorporate the principles of good assessment identified earlier, identify the key considerations that need to be addressed in producing a well constituted assessment plan:

- The assessment plan specifies the assessment methods to be used, their purpose, the marks to be allocated, and the timing of assessments.
- The selected assessment methods are valid for assessing the knowledge, skills and attitudinal components specified, and at the appropriate levels.
- The assessment methods are well constructed and sufficiently varied to enable learners to display understanding/competence through different mediums.
- The assessment methods are planned to make effective use of time and resources in producing sufficiency of evidence.
- The assessment methods provide fair and reliable assessment opportunities.
- The key aspects of the assessment plan are explained to learners.
- Opportunities are provided for learners to seek clarification on assessment requirements.
- Ways to ensure the authenticity of assessment evidence are identified.
- The assessment plan is reviewed at agreed times and up-dated as necessary.

Of note, it is important to recognize that in practice there is often a trade-off in terms of meeting the various principles of good assessment. For example, sufficiency is typically problematic in that what constitutes a sufficient range of assessment evidence (derived from a variety of methods) and how much evidence is sufficient is open to judgement. Indeed, Knight (2006) has pointed out that reliable judgements can only be made where there have been several observations from multiple observers in a range of contexts, which is not very practical in terms of resources.

(b) Judging evidence and making assessment decisions

Being an assessor is in many ways akin to being a 'caring' detective. The assessor is responsible for ensuring the quality of assessment (e.g., checking that assessment practices are based on principles of good assessment, as identified above). Part of this quality relates to principles of fairness and flexibility, which are to ensure that students are not discriminated against and the assessment process and expectations are clearly communicated and transparent. Also, where appropriate (without compromising the standard), flexibility can be used to accommodate individual student's special needs.

However, the most significant challenge in making valid assessment decisions revolves around considerations of appropriate standard or level of proficiency. While we would all like to have clear standards from which to base assessment decisions, this is often difficult to achieve in practice. Certainly, the explicit and valid identification of performance criteria is important here. Failure to appropriately make explicit the key constructs/elements that underpin the performance areas will seriously undermine the validity of assessment. Unfortunately, even when criteria are well derived and delineated, actual judgment in terms of how well a particular student performs requires interpretation, and this can vary across markers. Much of the problem is identified by Knight (2006) who argues that complex learning cannot be reduced to something simple enough to measure reliably: the more complex the learning, the more we draw on connoisseurship (Eisner 1985) rather than measurement to make our judgment (p. 38).

Indeed, attempting to break down standards into highly detailed and specific criteria actually results in a level of reductionism that both fails to capture the holistic contextualized performance as well as encouraging students to focus on these more atomistic components, mitigating the development of understanding and transfer.

However, despite the problems outlined above, there are ways to mitigate their impact and we have a responsibility to address these as best we can. Apart from accountability concerns, we are making decisions about our students that can have profound influence on their life opportunities, not least in the field of future educational access and employability.

A final important consideration in this context relates to the authenticity of assessment evidence provided by students, especially with the availability of resources on the internet and more assessment being conducted outside of the traditional exam format. In the traditional exam format, we could be reasonably sure that what was produced by individual students was their own work. However, in today's globally wired world, plagiarism is a serious assessment concern.

Bearing in mind the considerations identified above, the following criteria identify the key areas of practice for making the best judgments we can:

- Learners are provided with clear access to assessment.
- The assessment evidence is judged accurately against the agreed assessment criteria.

- Only the criteria specified for the assessment are used to judge assessment evidence.
- The assessment decisions are based on all relevant assessment evidence available.
- Inconsistencies in assessment evidence are clarified and resolved.
- The requirements to ensure authenticity are maintained.

(C) Providing feedback on assessment decisions
The importance of feedback is fundamental to the learning process as documented earlier—especially for formative assessment. As Gibbs (2008) highlights:

> Research in schools has identified that the way that teachers provide and use feedback, and engage students with feedback, makes more difference to student performance than anything else that they can do in the classroom. (p. 6)

It is also important to ensure sound recording, collation and security procedures in conducting assessment. The following criteria identify the key areas of practice for providing feedback and securing assessment outcomes:

- The assessment decisions are promptly communicated to learners.
- Feedback to learners is clear, constructive and seeks to promote future learning.
- Learners are encouraged to seek clarification and advice.
- The assessment decisions are appropriately recorded to meet verification requirements.
- Records are legible, accurate, stored securely and promptly passed to the next stage of the recording/certification process.

4.2.2 Strategies to Enhance Assessment Practices

Firstly, it is important to be clear that whatever strategies are employed, the actual design of assessment items and conduct of assessment activities must be congruent with the principles and standards documented earlier. Secondly, the CDIO standard is not prescriptive about the use of specific assessment methods; rather it advocates an eclectic approach in which methods should be closely calibrated to the types of learning outcomes being assessed. However, the emphasis on real world engineering projects and tasks, as well as the integration of skills across subject and domain fields, requires the refocusing of assessment towards a more performance-based rather than pencil and paper tests. While students will need to demonstrate key knowledge bases and understanding of conceptual knowledge in written and other 'paper and pencil' tests, this needs to be sufficiently augmented by integrated real world projects and learning experiences (e.g., Capstone Projects, Design Implement Experiences).

All assessment methods have strengths and limitations in terms of the types of assessment evidence they can generate and their usefulness will be largely dependent on the learning outcomes being assessed. For example, while multiple-choice items can be very effective and efficient for assessing knowledge and understanding (e.g., specific types of thinking such as analysis, comparison and contrast, evaluation, inference and interpretation), they have little validity for assessing integrated skills in complex problem solving activities. Similarly, performance-based items, which are perhaps the most valid for assessing the more complex real-world performances emphasized by the CDIO curriculum, are much more time and resource consuming and provide significant challenges to reliability. How to develop assessment systems for performance tasks will be considered in some detail later in the chapter.

Within this context, it is suggested that combinations of the following strategies can contribute to the effectiveness and efficiency of assessment practices in given assessment situations. The strategies are not meant to be exhaustive or summative, but represent practical frames from which assessment decisions can be thoughtfully made and practically customized to the particular situated assessment context:

(a) Produce assessment activities that are interesting and challenging
(b) Integrate a range of learning outcomes in assessment activities
(c) Provide as much transparency as possible in the assessment process
(d) Utilize student collaboration in formative assessment.

(a) Produce assessment activities that are interesting and challenging

One of the central themes of CDIO is to make engineering more real world, practice-based and interesting for students. In Chap. 3 a pedagogic framework was outlined and illustrated as foundational to the CDIO approach. However, a well constituted syllabus is of limited value in the hands of faculty who lack competence in pedagogic practices and the ability to create interesting and engaging learning experiences. Similarly, if our assessment activities lack interest for students and encourage rote learning, we should not be surprised to see them adopt the kind of 'surface approaches' to learning as documented by Marton (1984). In these situations students will learn what is necessary for the purposes of assessment, but are unlikely to derive both a real understanding of the subject or a genuine interest in it. Once the assessment process is finished, much of what was learned will soon be forgotten. In contrast, where students find the assessment activities interesting and sufficiently challenging, they are more likely to develop a genuine interest in the learning involved (Struyven et al. 2002). It is motivation for mastering the tasks set that leads to a desire for understanding the important concepts and principles of a subject and makes possible the transfer of learning (McTighe and Wiggins 2000).

(b) Integrate a range of learning outcomes in assessment activities

There is an old English saying, "kill two birds with one stone". In the context of assessment, this means getting the best efficiency from assessment activities and situations. The more assessment activities enable coverage of a range of learning outcomes, especially if they integrate topic and domain areas, the greater the potential benefit in terms of both student learning and assessment efficiency.

Furthermore, for CDIO skills such as Personal and Professional Skills and Attitudes, integration is important for effective learning. As documented in Chap. 2, the development of good thinking most effectively occurs within the context of domain specific knowledge. Similarly, where issues of a value-laden and ethical nature are involved, there is a need for both learning and assessment to be contextualized to the subject domain. Teaching ethics and values separate from real life contexts is similar to teaching thinking without reference to a knowledge domain. De-contextualized knowledge is difficult to transfer and may not even be perceived as meaningful by learners. Nucci's (2001) observation, in relation to the teaching of values, is relevant in this context:

> The greatest challenge for a teacher wishing to engage in domain appropriate practice is to identify issues within the regular academic curriculum that will generate discussion and reflection around a particular value. (pp. 178–79)

Students are firstly more likely to understand and internalize learning in the affective domain where it is contextualized to specific engineering contexts and issues. Secondly, in terms of assessment, there is more likely to be authentic assessment opportunities and greater validity in the assessment of such knowledge and skills. For example, in the world of engineering practices, as in any professional domain, there are ample opportunities to naturally infuse the full range of types of thinking as well as areas of ethical concern—both in terms of personal values and wider societal issues that involve professional ethics.

(c) Provide as much transparency as possible in the assessment process
Assessment should not be designed to mystify students about what they should be learning and to what standard. It is about developing the range of competences and proficiency levels that are deemed appropriate for the curriculum in question. When one goes for a driving test, there is full transparency as to what constitutes competent driving performance. It is also equally clear as to what gaps in knowledge and competence will lead to failing the test. Other things being constant, such as assessor reliability and fairness, etc., passing or failing will depend on the performances exhibited in the test situation.

Smart students have long worked out that the secret to success in assessment boils down to a basic logic: "know what needs to be learned, learn it and know that you have learned it". However, simply knowing this does not mean success either in learning or assessment outcomes. Within this simple maxim is an implicit added element called *effort*. Learning, at a high proficiency level typically involves considerable effort and time on task, which many students may not be prepared to do. Students who make such effort and develop such competences deserved to be successful—do they not?

It makes sense, therefore, for the reasons outlined above, to be transparent in the assessment process. This is consistent with the principle of fairness and provides students with a clear guide on what constitutes the standard for the course being studied. It is then their responsibility to develop personal autonomy in learning and make the necessary effort. Providing transparency in the assessment process can be achieved in the following two main ways:

- A syllabus in which the learning outcomes are clearly written and communicated to students
- Assessment criteria, calibrated to specific learning outcomes, which identifies the performance areas being assessed, the performance evidence that needs to be generated, and the performance criteria upon which this is based.

Bloxham and Boyd (2007) summarize this in terms of a module context:

> A number of researchers are now concluding that preparing students for assessment is not a distinct stage in a module but should be part of an integrated cycle of guidance and feedback, involving students in active ways at all stages. (p. 71)

Another useful means for promoting transparency is through the use of well constituted assessment rubrics, especially for performance-tasks. These provide both students and assessing faculty with clear descriptions of performance at various levels. The use of assessment rubrics will be outlined further in the chapter.

(d) Utilize student collaboration in formative assessment

As assessment is fundamentally linked to learning and teaching, it makes sense to utilize the main stakeholders (faculty and students) to collaboratively make this 'system' work to the best advantages of both. For example, from a student's perspective, the ideal would be to use assessment to optimally support the learning process through the various processes of formative assessment. Similarly, from a faculty perspective, we would like to be able to identify student learning concerns/problems quickly and be able to effectively and efficiently deal with them—whether through instructional design or other learning support means. How then, might we create the kind of symbiosis that makes possible the best collaboration between faculty and students, without compromising the quality and credibility of final summative assessments?

The first and most fundamental way to utilize this collaboration is to help students to develop their own self-assessment capability. Students who are able, in large part, to identify what they know and don't know are already a long way to becoming independent learners and taking a significant load off faculty in terms of instructional and remediation time. Time spent in developing student's capability to self-assess will result in better learning for students as well as making the instructional process more efficient.

Secondly, having students involved in peer assessment further supports student learning of the subject content, assessment skills, and facilitating collaborative learning as they act as tutors and mentors for each other. While this needs to be carefully monitored and supported by faculty it will, over time, take considerable resource pressure off them.

4.2.3 Develop Effective and Practical Assessment Instruments

Firstly, good design is fundamental to all assessment items, whether fixed response, essay type or performance-tests. This is well documented in the literature (Haladyna 1997; Osterlind 1989). However, different assessment items, apart from offering different assessment evidence on students' learning, also provide different challenges in terms of making assessment decisions. For example, while multiple-choice items require skill in design, and the production of a large bank of useful items is time consuming, marking is easy and efficient. Open response (essay-type) and performance-based items, in contrast, require a more elaborate marking system and are prone to subjectivity.

However, performance based assessments are potentially the most valid forms of assessment as they provide assessment opportunities where students can display key competences in real world or simulated activities. Also, as Tombari and Borich (1999) explain:

> A performance assessment is a test that tries to determine if a learner "really knows" about something, or has deep understanding. It does this by challenging learners with tasks that ask them not simply to recall knowledge but to construct or organize it, not just to solve problems but to demonstrate a disciplined approach requiring strategic thinking and metacognition. (p. 148)

Performance-based assessment is most akin to the CDIO assessment approach and offers the following assessment advantages over more traditional 'pencil and paper' based approaches:

- Offers greater validity as the focus is on real life performance.
- Measures a range of complex skills and processes in real world or authentically simulated contexts.
- Links clearly with learning and instruction in a planned developmental manner.
- Motivates students through meaningful and challenging activities.

Invariably, performance-based assessment poses challenges for faculty in terms of:

- More time consuming than 'paper and pencil' type assessment.
- Where courses focus on underpinning knowledge, there is less opportunity for performance-based assessment.
- As these items often involve professional judgement, there is always the problem of subjectivity in marking.

The issue of subjectivity in marking is indeed a serious one, especially in project activities that integrate a range of competency areas across modules and subject domains. Bloxham and Boyd's (2007) observation captures this problem precisely:

> …the research suggests that providing fairness, consistency and reliability in marking is a significant challenge caused by the inherent difficulty of reliably marking complex and subjective material combined with our own marking dispositions. (p. 87)

The actual design process for performance-based learning tasks was outlined and illustrated in Chap. 3. Such tasks can encompass a wide range of activities, for example:

- Real work projects and tasks
- Simulations
- Problem solving through case studies
- Presentations
- Any activity that essentially models what would be done by professionals in the world of work.

While CDIO is not prescriptive in terms of specifying what marking systems are to be employed, the development of marking systems for performance-based learning tasks should carefully address the following assessment considerations:

- Assessment areas
- Performance criteria
- Assessment evidence
- Assessment rubrics.

Assessment Areas

Assessment areas constitute the main *performances* that are to be assessed in any performance-based activity, and typically a number of assessment areas can be validly assessed. For example, in a project-based activity, there are usually opportunities to assess critical and creative thinking, teamwork and communication, ethical issues, as well as the technical engineering content areas. However, just because a performance test offers such opportunities for assessment, this does not automatically mean that all possible performance areas must be assessed, especially in summative terms. What is assessed from such an activity should be considered in relation to other assessment components for the module or unit of study. For example, if an area has been sufficiently assessed elsewhere, it may be more practical to assess other important areas that have not so far been assessed. However, it is of course useful, whenever feasible, to provide appropriate formative assessment in all the significant performance areas.

Once the summative assessment areas have been identified for the learning activity, it is then necessary to identify the marks allocation or weighting for each of the designated areas. This should reflect the learning outcomes and their relative importance within the module or unit context, as well in relation to other components of the assessment plan. An example for a design-implement project is identified in Table 4.1.

Table 4.1 Assessment components for a Design-Implement Project

Assessment components	Mark weighting in (%)
Production of car chassis components (based on practical work done in the workshop)	30
Assembly of car chassis components (based on practical work done in the workshop)	10
Performance of the model F1 car in the racing Challenge (speed and stability under test conditions)	20
Teamwork (e.g., goal setting, management of team-roles and responsibilities, dealing with conflict/challenges, etc.)	30
Oral presentation (e.g., organization, clarity, and effectiveness of oral communication)	10
Total	100

Performance Criteria

Performance criteria are the more specific and measurable elements/behaviours that underpin the wider performance to be assessed in an assessment area. For example, in assessing 'Demonstrate Effective Written Communication' (a component of Interpersonal Skills, Communication) the performance criteria may include 'write with logical organization and clear language flow', 'use concise and precise language', 'use correct grammar, spelling and punctuation', etc.

In the case of SP, the customized syllabus, written to include specific learning outcomes for each of the CDIO Skills, has provided the necessary guidance for enabling alignment between the learning outcomes and the performance criteria for assessment. The important point is to look carefully at the learning outcomes in each of the chosen CDIO skill areas and ask the question:

> Will this performance task offer the student a realistic opportunity to demonstrate that he/she can meet this learning outcome?

For example, Table 4.2 shows how one school chose to organize their marking system for a third-year capstone project. From an analysis of the range of activities that the students were going to engage in over the course of the project, and in relation to the customized SP syllabus, they derived the following performance criteria that would form the key focus for the assessment areas.

Assessment Evidence

Assessment Evidence refers to the range of performances and products that can be validly and efficiently considered in making an assessment decision. In making assessment decisions, it is firstly necessary to consider the range of evidence that can be generated by the various activities in relation to the performance areas and criteria. For example, in assessing teamwork, a wide range of evidence sources can

Table 4.2 Raw mark form—projects with physical deliverables

PROJECT NO. ____________		NAME & ADM NUMBER OF STUDENTS
ASSESSMENT AREAS	PERFORMANCE CRITERIA	
1. Conceiving (Default 15 %) [Range:10 – 25 %] Selected Weightage: ____	1.1 Eliciting market needs and opportunities 1.2 Defining functions and concepts of the system 1.3 Modelling system to verify goals 1.4 Development of project plan	
2. Designing (Default 25 %) [Range:15 –35 %] Selected Weightage: ____	2.1 Formulation of design plan 2.2 Selection of final design 2.3 Consideration of project costs 2.4 Evaluation of selected design	
3. Implementing (Default 15 %) [Range:5 – 25 %] Selected Weightage: ____	3.1 Designing the implementation process 3.2 Planning for hardware (or software) realisation 3.3 Testing, verifying, validating and certifying	
4. Operating (Default 5 %) [Range:5 – 10 %] Selected Weightage: ____	4.1 Planning training and operating procedures 4.2 Suggesting improvements to project 4.3 Planning for project disposal	
5. Teamwork (10 %)	5.1 Identification of goals and work agendas 5.2 Utilisation of team strengths 5.3 Application of ground rules and management of conflict	
6. Effective Communication (15 %)	6.1 Logical organisation of content and language flow in the project report 6.2 Using correct language and grammar in the project report 6.3 Producing engineering drawings 6.4 Using effective oral communication	

(continued)

Table 4.2 (continued)

PROJECT NO. ______		NAME & ADM NUMBER OF STUDENTS
ASSESSMENT AREAS	PERFORMANCE CRITERIA	
7. Personal and Professional Skills & Attributes (15 %)	7.1 Using a range of critical and creative thinking skills 7.2 Monitoring and reviewing quality of own thinking 7.3 Managing learning 7.4 Acting in a manner consistent with professional codes and ethics	
CDIO skills 80 % (For the above raw marks, items 1–4 = 60 %, items 5–7 = 40 %) Exhibition 10 % (Duty/Exhibited = 5, Bronze = 6, Silver = 8, Gold = 10) Deadline 10 % (Project Deadline —as per page 6 of logbook) Marks ratio between supervisor and co-examiner = 2:1		Score 5 to 1 in the boxes above. Key: 5 = Consistently met to a very high standard 4 = Mainly met to a high standard 3 = Mainly met to an acceptable standard 2 = Partially met to an acceptable standard 1 = Very poor performance

be generated and used to make a valid assessment decision. These could include the following:

- Feedback from students (e.g., peer assessment)
- Lecturer observation of student interactions
- Questioning
- Meeting deadlines and objectives
- Student logs/journals

For SP projects, the following generic types of assessment evidence are typically produced:

- Reports
- Progress reviews
- Logbook
- Scheduling documentation
- Engineering drawings
- Artifacts (e.g., models, prototypes, programmes, operating manuals)
- Presentations
- Responses to questions (oral, written)

Invariably, the greater the range of evidence sources that can be accessed (providing they are sufficiently valid and reliable), the more likely it is that we can make accurate assessments of performance.

Assessment Rubrics

Assessment Rubrics are rating scales in which a prepared scoring system is used for assessing learner performance for a particular task or assessment area across a number of levels of that performance (usually 1–5, in which 1 denotes a very poor performance and 5 denotes a very good performance). Assessment rubrics are most useful when assessing more complex activities where assessment of a performance is often one of variation of performance, involving a high level of inference. For example, in assessing teamwork, it is often not a clear case of being either effective or ineffective in this performance area, but rather variation along a continuum from very effective to very ineffective. Furthermore, as there are many aspects and potentially different interpretations of what constitutes effective teamwork, it is open to different inferences by different assessors. The extent to which assessors are likely to differ in terms of an assessment decisions relating to an area of performance determines the level of inference.

In performances in which all assessors, assuming expertise in the area, would largely agree on the level of performance, we can say that assessment is low inference. This would be the case in most procedural aspects of a performance in which there are clear and established, almost algorithmic, standards relating to effective performance. In these assessment situations a checklist is a more appropriate tool and easier to use marking system. However, in areas such as creativity and aesthetics, while there are a number of features that can be identified as criteria of quality, there is still a high level of subjectivity in terms of personal interpretation of what this looks like overall. This would represent a high inference assessment, where assessors may have quite diverse perceptions of what is good and poor performance. For assessing these areas, the descriptors of different levels of performance are useful in mitigating the variation of assessment decisions and enhancing reliability.

In using rubrics, decisions need to be made on whether to assess more holistically or analytically in relation to performance areas and criteria. Essentially this relates to whether to assess the performance area overall, e.g., oral communication and give a score; or break it down into key components/constructs, score these individually, then derive the overall score. There are merits in both approaches (Biggs 2003; Gosling and Moon 2003). Holistic rubrics enable a focus on the overall performance and are more economical in terms of assessment time. They are typically used for summative assessment and where some variation in reliability in parts of the assessment components can be accepted, provided the overall assessment decision has good validity and reliability. In contrast, analytical rubrics enable a much greater focus on the specific elements of the areas of learning involved and make possible a much better utilization of formative assessment in the assessment process. One type of rubric is not inherently better than another (Montgomery 2001); it really depends on the assessment purpose and context in which rubrics are used. In SP, it is not mandated that staff use rubrics, but they are encouraged to do some in cases of moderate to high inference assessment.

It is also important to remember that the rubric *does not* make the assessment decision; this is the responsibility of the assessing lecturer(s). Rubrics provides a guiding frame for focusing attention on the key elements/constructs (performance criteria) of the assessment area and summary descriptors of a range of performances. Designing effective and efficient rubrics can be a difficult and frustrating activity for faculty not familiar with such assessment tools. It is essential that training and support is provided by experienced educational development faculty well versed in rubric design. In our experience, educational development faculty working collaboratively with school-based faculty has proved most productive in terms of acceptance and ownership of the scoring systems developed.

There are a number of established texts on how to construct various rubrics (Butler and McMunn 2006; Stiggins et al 2006). Some of the more salient considerations are summarized in Table 4.3.

In scoring student performance, it is often the case that some students do not nicely 'fit' all the behavioural indicators in any one description of performance

Table 4.3 Key considerations in rubrics design

Rubric design		
Identifying and writing criteria	Writing descriptions of performance:	Guide to scoring performance:
– Criteria identify the most Important constructs/ elements of the performance being assessed	– Descriptors clear and concise – Descriptors use language that is familiar and understandable by assessors and students being assessed	– Performance areas and criteria are differentiated in the weighting of marks allocated where appropriate (e.g., importance, complexity, etc.) – The scoring system makes clear how rubric scores are translated into grades
– Criteria are clearly aligned to the designated learning outcomes for the performance area	– Descriptors provide accurate descriptions of the performance at the designated level – Qualitative terms (e.g., Many, some, few, etc.) need to be clarified and understood by assessors and students	
– Criteria are explicitly stated and measurable based on the evidence that can be generated by the assessment method(s) employed		

(e.g., they may fit most indicators quite well but are better or worse on the others). In this situation, it is practical to choose the description that you feel is the most appropriate in terms of the score to be given for that performance area. This can be moderated and/or adjusted holistically at the end of the assessment process for the task (especially in borderline cases). Our experience clearly informs us that there is little point in having highly detailed marking systems that are burdensome in practice. As a consequence, for purposes of summative assessment, we have used a holistic rubric format, as illustrated in Table 4.4, which shows a typical rubric design that has been used in a range of contexts for scoring performance in Oral Presentation Skills.

This is a fairly standard rubric design in which a performance area (in this case oral communication) is broken down into key behavioural indicators relating to the SP customized syllabus. These become the basis for the five levels of descriptive performance. Marks can then be allocated to these broad band descriptors in terms of wider assessment weightings and grading formats. Converting rubric scores into grades is more a question of logic than any particular mathematical formula. What is essential is that the marks allocation calibrate to what has been decided in the overall assessment plan for the module or unit of study.

Table 4.4 Rubric Template for Oral Presentation Skills

Scoring rubric for oral presentation skills

The scoring rubric provides descriptions of five levels of student performance relating to Oral Presentation (where a score of 5 represents very good performance and a score of 1 represents very poor performance).

The rubric is underpinned by specific behavioural indicators of oral presentation, these are:

- Clarity of voice, tone and modularity
- Appropriateness of presentation structure and style to specific audience
- Calibration of non-verbal communication to the spoken words (e.g., posture, eye contact and gestures)
- Answering questions in a clear, concise and focused manner.

In scoring student performance, it is often the case that some students do not nicely relate to all the behavioural indicators in any one description of performance (e.g., they may fit most indicators quite well but are better or worse on the others). However, choose the description that you feel is the most appropriate in terms of the score to be given for the individual student.

Score	Description of performance
5	Voice is consistently clear and effectiveness in terms of tone and modularity Presentation structure and style fully relates to audience Non-verbal communication is highly calibrated to spoken word All questions answered in a clear, concise and focused manner
4	Voice is generally clear and effectiveness in terms of tone and modularity Presentation structure and style mainly relates to audience Non-verbal communication is calibrated to spoken word Most questions answered in a clear,concise and focused manner
3	Voice is occasionally clear and effectiveness in terms of tone and modularity Presentation structure and style relates to audience in part Non-verbal communication is sometimes calibrated to spoken word Some questions answered in a clear, concise and focused manner
2	Voice has limited clarity and effectiveness in terms of tone and modularity Presentation structure and style rarely relates to audience Non -verbal communication is mainly not calibrated to spoken word Few questions answered in a clear, concise and focused manner
1	A very poor performance in this area of competence

4.3 Summary

This chapter has outlined the CDIO framework for assessment, and documented how we have customized it in practical ways to the SP context. Most significantly, we have attempted to continue in the vein of the 'science of learning' approach outlined in the previous chapter. While assessment is not an exact science, much is known about good assessment practices in terms of the principles and standards documented in this chapter. Furthermore, the CDIO focus on 'Learning Assessment' is well validated in the literature and is part of good assessment practice. The main challenge is to encourage and support faculty in the design of robust assessment systems and items that are valid and reliable, in the context described, as well as feasible in the real world of assessment practice. We recognize the importance of professional judgement and we must trust staff to use it in making assessment decisions.

References

Banta TW, Jones EA, and Black KE (2009) Designing effective assessment: principles and profiles of good practice. Jossey-Bass, San Francisco
Biggs JB (1996) Enhancing teaching through constructive alignment. High Educ 32(3):347–364
Biggs JB (2003) Teaching for quality learning at university. Open University Press/Society for Research into Higher Education, Buckingham
Bloxham S, Boyd P (2007) Developing effective assessment in higher education. Open University Press, Maidenhead
Boud D (1995) Ensuring assessment fosters learning: meeting the challenges of more students with diverse backgrounds. In: Armstrong J, Conrad L (eds) Subject evaluation. Griffith Institute for Higher Education, Queensland
Boud D (ed) (1988) Developing student autonomy in learning. Kogan Page, London
Boud D (2000) Sustainable assessment; rethinking assessment for the learning society. Stud Continuing educ 22(2):151–167
Butler SM, McMunn ND (2006) A teacher's guide to classroom assessment: understanding and using assessment to improve student learning. Jossey-Bass, San Francisco
Criteria (3rd edn). SEEC, London
Eisner EW (1985) The art of educational evaluation: a personal view. The Falmer Press, Basingstoke
Gosling D, Moon J (2003) How to use learning outcomes and assessment
Gibbs G (2008) How assessment can support or undermine learning, in TLHE, frontiers in higher education. CDLT, Singapore
Gray P (2007) Student learning assessment. In: Crawley E, Malmqvist J, Ostlund S, Brodeur D (eds) (2007), Rethinking engineering education. Springer, New York
Haladyna TM (1997) Writing test items to assess higher order thinking. Allyn & bacon, Needham Heights
Knight PT (ed) (1995) Assessment for learning in higher education. Kogan Page, London
Knight PT (2006) The local practices of assessment. Assess Eval High Educ 31(4):435–452
Marton F (1984) The experience of learning. Scottish Academic Press, Edinburgh
McTighe J, Wiggins G (2000) understanding by design. Association for supervision and curriculum development
Montgomery K (2001) Authentic assessment: A guide for elementary teachers. Longman, New York
Nucci IP (2001) Education in the moral domain. Cambridge University Press, Cambridge
Osterlind SJ (1989) Constructing test items: multiple-choice, constructed response, performance and other formats. Kluwer Academic Publishers, Dordrehct, Netherlands
Prosser M, Trigwell K (1998) Understanding learning and teaching: the experience in higher education. Open University Press, Buckingham
Perkins DN (1992) Smart schools. The Free Press, London
Ramsden P (1992) Learning to teach in higher education. Routledge, London
Rowntree D (1987) Assessing students: how shall we know them. Harper Row, London
Stiggins R, Arter J, Chappuis J, Chappuis S (2006) Classroom assessment for student llearning. Pearson, New Jersey

Struyven K, Dochy F and Janssens S (2002) 'Students' perceptions about assessment in higher education: a review'. Paper presented at the joint Northumbria/EARLI SG assessment and evalaution conference: learning communities and assessment cultures, University of Northumbria, 28–30 Aug

Tombari M, Borich G (1999) Authentic assessment in the classroom. Prentice-Hall, New Jersey

Chapter 5
Using Information-Communication Technologies to Support Learning and Teaching

5.1 Introduction

> With good pedagogy as the guiding goal, technologies can be employed selectively and sensitively to make a distinct contribution to teaching and learning. (Perkins 1995, xvi)

This chapter could easily have been written without any reference to CDIO. The content and the approach taken are essentially generic to the use of information-communication technologies (ICT) for enhancing learning effectiveness and efficiency in any educational context. However, ICT's are now so much part of the educational landscape, I felt it important to provide a critical and practical frame on their usage within the content of this book. Perhaps, even more significantly, while ICT have been around for some time now, and are continually and rapidly evolving, now is the ideal time for mainstream faculty to really exploit their affordances.

I must confess to previously being sceptical concerning the often muted significant benefits of ICT in enhancing learning effectiveness—at least in the short term anyway. Like many others, I regularly experienced frustration when using technology-based databases and software, often thinking "why is it that such a simple process seems like the Mars mission?" Also, I remember, in the early days of the technology euphoria, waiting to listen to a keynote talk at a conference on the supposed benefits of technology in teaching—except that the speaker was never able to get his work (not even his PowerPoint slides) up on the screen. He had to abort the presentation. Such experiences did little to inspire us non-techie folk to embrace technology for learning in any sustained manner.

Sallis and Hingley (1991) once referred to education "as a creature of fashion" (p. 9). That metaphor is perhaps most apt in the field of online or e-learning. It does not seem so long ago that e-learning was being touted, to use an old English metaphor, "as the best thing since sliced bread". However, more recently, we hear references to e-learning as a "Thwarted Innovation" (Zemsky and Massy 2004). Similarly, Hargreaves and Fullan (1998) concluded:

D. Sale, *The Challenge of Reframing Engineering Education*,
DOI: 10.1007/978-981-4560-29-0_5, © Springer Science+Business Media Singapore 2014

> By far the biggest weakness in how schools use new technologies are pedagogical and strategic. (p. 84)

More recently, Oliver et al. (2007), commenting on the widespread availability of ICT to create engaging and effective learning settings noted that:

> What appears to be still missing for teachers is appropriate guidance on the effective pedagogical practice needed to support such activities. (p. 64)

The reflections by Shea—Schultz and Fogarty (2002) provide a very interesting insight into this apparent e-learning paradox:

> One thing is certain—e-learning will evolve into something so simple, so elegant yet all persuasive and natural, that our grandchildren will wonder in dismay why we didn't see it coming. (p. 165)

> Truly human-friendly technological design won't appear anytime soon. Computer, networking and software engineers cast the die five decades ago. (p. 89)

Fortunately, the above scenario is changing for the better. Firstly, the technologies are becoming more stable, faster (e.g., video streaming) and rapid development software tools (e.g., SoftChalk Lesson Builder, Camtasia Studio, VoiceThread) now require little learning time for high affordance practical application.

Secondly, and perhaps most significantly, there is a refocusing of the use of ICT towards a pedagogic perspective rather than technological use per se. Instead of technology being used for technology sake, we are asking more critical questions relating to learning effectiveness and efficiency. For example, if the use of technology does not result in a significant gain (however defined) in either learning effectiveness or efficiency for a group of learners in a particular context, then why use it?

In the following sections I will outline a framework from which ICT can be used effectively and efficiently across the full range of teaching and learning modes (e.g., face-to-face, fully online and blended). It is based on the pedagogic framework presented in Chap. 3 in which core principles of learning were identified and explained in relation to aspects of the learning process. These constitute the foundation of a pedagogic literacy, which applies to the design of learning experiences irrespective of delivery mode or context. There are many affordances of ICT when used from a pedagogic perspective framework, and in the context of user-friendly rapid development software and collaborative web communication tools, the possibilities are indeed exciting.

5.2 A Design Framework

The development of an effective and efficient learning design (irrespective of delivery mode) can best be achieved through recognition and application of the following:

1. Good learning design is always grounded on core principles of learning
2. ICT are used to enhance specific aspects of the learning process
3. The completed blended learning design maximizes the affordances of a range of learning modes and mediums.

1. Good learning design is always grounded on core principles of learning

As mentioned earlier, there is now an increasing focus on the pedagogy rather than the technology in the use of ICT. Schrage (1990) is spot on when he argues that:

> Technology is really a medium for creating productive environments. (p. 67)

Similarly in discussing e-learning, Horton (2006) makes the important point that:

> At its best, e-learning is as good as the best classroom learning. At its worst, it is as bad as the worst classroom learning. The difference is design (p. 3).

The core principles of learning are equally applicable to designing and managing learning in the online or blended learning environment as in the face-to-face context. Learning online does not change the way the human brain functions or the basic processes of learning. ICT, in the present context anyway, do not fundamentally affect human cognitive functioning and therefore key aspects of the way humans learns. There is no 'learning revolution' in terms of the way people learn; rather there is an information and processing revolution. Human brains are no different now than 50 years ago and, if the anthropological evidence is correct, maybe not even 50,000 years ago. While many scientists may presently be seeking to integrate research findings from genetics, robotics, information technology and nanotechnology to improve human psychological functioning, we are probably dealing with the brain as it is for at least the next 20 years. What this means is that we will still need to design our instructional strategies to be most consistent with existing cognitive functioning and other nuances of the human condition. Clarke and Lyons (2005) analysis is still relevant for the foreseeable future:

> The most robust instructional principles are those based on a model of human psychological learning processes….Any given instructional method will be effective or ineffective depending on the extent to which it supports or disrupts basic-learning psychological processes regardless of the delivery media. (p. 594)

Quite simply, disorganized and over complex content in the online environment is no less disruptive than in the face-to-face context—perhaps even more so. Similarly, dull is dull, wherever, whenever.

Also, within this frame, it is important to consider what human brains are both capable and disposed to do well, in relation to information processing, as compared to ICT. This provides significant guidance as to the learning contexts in which technologies may be most effective. For example, we know ICT are much better than the human brain at:

- Retrieving information from vast resource banks of data
- Rapidly, accurately and effectively processing complex sequences of clearly defined facts
- Reconstructing and re-presenting large amounts of information.

Hence, in terms of providing access to knowledge bases and sources, ICT offer almost unlimited opportunities. Vast amounts of information are contained in CD Roms, databases, the Internet and through communication tools such as email, bulletin boards and discussion forums. The capacity to decentralise the structure of knowledge bases and reconstruct them in dynamic customised digestible bits (knowledge warehousing) makes knowledge even more directly accessible. Furthermore, packages that present knowledge in organized formats, which enable learners to engage in constant recall practice with immediate feedback can facilitate knowledge acquisition.

In contrast, the human brain is better (at present anyway) than the computer at:

- Conceptualising ambiguous problems
- Exploring concepts
- Formulating and communicating ideas.

This would seem to be strongly supported by Brown's (2009) extensive experience in working with creative design teams. He concluded that:

> The internet helps move information around but has done little to bring people together. Creative teams need to be able to share their thoughts not only verbally but visually and physically as well. I am not at my best writing memos. Instead, put in a room where somebody is sketching on a whiteboard, a couple of others are writing Post-its or sticking polaroid photos on the wall, and somebody is sitting on the floor putting together a quick prototype. I haven't heard of a remote collaboration tool that can substitute for the give-and-take of sharing ideas in real time. (p. 30)

The extent to which ICT can enhance the quality of students thinking, especially creative thinking, is of key interest from a pedagogical point of view. The importance of thinking in effective learning is well documented (e.g., Marzano 1992; Swartz and Parks 1994). Furthermore, the development of good thinking has been a challenge to traditional forms of instruction long before the availability of computer based technologies. In order to ascertain the potential usage of IT in promoting types of thinking it is necessary to identify the instructional features and conditions that are conducive to the promotion of such thinking. It is firmly established that effective student thinking can be encouraged through:

- Active involvement in the learning process (questioning, discussion, debates, etc.)
- Engagement in real world problem solving tasks (projects, case studies, etc.)
- Collaborative team-learning (group projects, discussion forums, etc.).

There are now many ICT applications that can be used to facilitate and enhance thinking and learning in such areas. These include:

- Online tutorials involving active problem-solving with feedback
- Hypermedia software integrating knowledge, multimedia, activities and feedback
- A range of communication tools (e.g., email, blogs, bulletin boards, forums, etc.)
- Constructing software (e.g., desktop publishing, spread sheets, etc.) where learners can produce, manipulate and change information
- Simulations and virtual reality programmes.

From a pedagogical point of view, we need to be clear about the types of thinking that we are trying to promote and provide practice in, as documented in some detail in Chap. 2. From this standpoint we can ascertain how certain technologies and their particular use may contribute to enhancing the development of such skills. However, it is necessary to recognize that the technologies themselves do not ensure good thinking. As Melchior (1997) points out:

> One pervasive myth is that the technologies themselves teach important complex skills…they need to be identified, taught, modelled, and reinforced by capable teachers. (p. 91)

For example, in a chemical engineering module, using dynamic simulation to promote critical thinking, Sale and Cheah (2011) noted that it was the design of activities, and specific questioning, cueing such thinking skills as analysis, comparison and contrast, evaluation and making inferences and interpretations, that was most significant in promoting good thinking. Where activities were not challenging, the simulator lost this capability for enhancing the development of such critical thinking skills. It was also noted that in situations where both faculty and students had shared notions of what constitutes good thinking, there was evidence of further enhancements in the quality of student thinking.

Furthermore, and particularly important in the context of promoting good thinking, there is real concern over the validity and usefulness of much internet content. Indeed, Keen (2007) makes a damning criticism of so-called internet expertise knowledge contained in such sites as Wikipedia:

> …the real consequence of the Web 2.0 revolution is less culture, less reliable news, and a chaos of useless information. One chilling reality in this brave new digital epoch is the blurring, obfuscation, and even disappearance of truth. (p. 16)

Sylwester's (1995) summary, from extensive review of the literature, pulls the present discussion together nicely:

Our brain is better than a computer at conceptualising ambiguous problems. Conversely, a computer is better at rapidly, accurately and effectively processing complex sequences of clearly defined facts (p. 120).

It makes logical sense, therefore, to use such knowledge when designing and managing e-learning, for example:

1. Model and morph what works well in face-to-face learning situations onto the online platform—where viable and cost-effective
2. Don't try to morph from the face-to-face what is not viable and cost-effective

3. Utilize the particular learning affordances (and there are many) that the online capability offers.

2. ICT are used to enhance specific aspects of the learning process

In this section, certain key affordances of ICT are identified and discussed in relation to pedagogic effectiveness and efficiency. The coverage is not intended to be comprehensive, and the potentialities are changing almost daily in this rapidly evolving field to claim a high degree of currency. However, working from a pedagogical perspective, the essential framing and analysis of ICT usefulness will remain current at the time of reading—so don't put the chapter down.

In designing learning experiences, it is important to identify the ways in which ICT (including online capabilities) can enhance aspects of the learning process. This firstly involves clear recognition of the generic unique affordances offered by the online environment and supporting technologies. These have been typically identified as:

- Anytime, anyplace access to online resources
- Hyperlinked multi-modal, dynamic content
- Global social networking.

Secondly—and this is where *pedagogic literacy* is of key importance—it is necessary to be able to identify the potential learning enhancement capabilities (and costs) of specific technology applications and technology combinations. The essential question then, goes something like this: "which e- tools can enhance specific aspects of the learning process, for what learners, how, and in what contexts, etc.?" In that an e-tool supports any of the core principles of learning, there are possible enhancements to aspects of the learning process. For example, perhaps the major affordance of the use of ICT is the capability to design multi-sensory learning experiences in the anytime, anyplace, hyperlinked online environment. People learn better when there is multi-sensory engagement, as identified in *Core Principle 6: Instructional methods and presentation mediums engage the range of human senses*. That's why good face-to-face teachers maximize their presentation style (e.g., voice modality, movement, eye contact, audio-visual aids, etc.) to gain best attention and engagement. The mainstream online learning environment does not, as to yet, provide such 'visceral reality' and neither can it viably compete with the virtual reality world of "Disney".

However, what the online environment offers is the capability to bring together a wide range of text based, multi-media and personnel resources way beyond what is possible in the traditional classroom through the simple but awesome power of the hyperlink. Hamilton and Zimmerman (2002) illustrate this vividly when they wrote:

> …the hyperlink, which is practicably without counterpart in the physical world of traditional academics. Within an internet document, hyperlinks are used to bring multisourced information into the primary text or to give the reader a path to alternative media. In essence, this eliminates the *physical* separation of material messages that are *logically*

> connected. In addition to text, hyperlinked messages may be pictures, sound files, animations, or video clips. External links can refer students to other information-rich Internet sites, including personal Web pages, specialized bibliographies, and professional specialists. (p. 270)

This provides the capability of creating networked resources that enable both faculty and students to create, share and continually devlop an extensive and varied range of resources that can support the desired learning outcomes. These can include:

- Centralized key resources relating to a module syllabus (e.g., learning guides, module maps, advanced organizers, annotated bibliographies of key resources, guidance on how to negotiate potential difficult topic areas)
- Selected prepared resources to support learning (e.g., notes, cases, videos, animations, activities)
- Selected web links to provide a networked architecture of extended and dynamic resources
- Access, where appropriate, to other digital learning exchange portals (e.g., libraries, specific learning communities).

Even from a more traditional teacher-centred instructional point of view, this capability provides faculty with a myriad of resource possibilities for facilitating desired learning outcomes to a range of learning groups and learner preferences. The ability to weave these resources into an effective—even creative—teaching strategy relates back to the earlier sections and considerations of good pedagogy. If teachers of yesteryear, armed only with a blackboard and different coloured chalks could still create interesting and engaging lessons—and there were those who could do this—then consider the creative potentialities in the present context.

Furthermore, it should be recognized that the use of asynchronous text can provide certain significant advantages over the typical face-to-face situation. In face-to-face learning there is often too much information to absorb and too little time for reflective thought and the kind of synergetic knowledge building that comes from good collaborative learning over time. Online text provides an opportunity to model such synergetic activity and help build deep learning (Marton et al. 1984). Hamilton and Zimmerman (2002) argue that the asynchronous text medium can create an excellent forum for thoughtful, extended, reflective dialogue. They highlight that:

> The medium supports iterative exchanges of information and opinions over an extended time period, so ideas are not merely "hatched" and delivered but rather allowed to evolve and be refined in a manner that makes information more convincing, narrative deliveries richer in detail, and learning more thorough. (p. 265)

A similar analysis can be applied to any other aspect of the learning process. For example, we know that good communication and rapport are fundamental to effective learning, and this is often seen as a potential strength of the face-to-face situation—at its best I might emphasize. However, we may be increasingly in situations where we might prefer to use a face-to-face mode of learning, but considerations of cost make this less viable. In the real world of educational provision, we may have to compromise our best learning offerings to the needs of cost efficiency. In these situations, we must apply our pedagogic skills to use available technologies in the most effective ways. In terms of communication and rapport building, apart from the clarity and organization of information, the ability to provide *Motivational and Attentional strategies.* (Core principle 3) and create *A psychological climate which is both success orientated and fun* (Core Principle 9), are particularly important. For example, without the personal resources of voice tone, gesture and body language, the online tutor has to use other communication skills and strategies to foster a welcoming, friendly and comfortable environment in which learners feel motivated to participate and collaborate. Hodges and Saba (2002) suggest that there are three role dimensions for online tutors to negotiate if they are to be effective in online tutoring:

- Organisational Role: This involves creating the agenda for the online programme, establishing objectives of the forum discussion, timetabling, creating procedural rules, and decision-making norms.
- Social Role: This involves creating a friendly social environment for learning. It will involve a frequent and lively presence, as well as a sense of humour.
- Intellectual Role: This is essentially about educational facilitation. As in any kind of teaching, the moderator should focus discussions on crucial points, ask pertinent questions, and probe responses to encourage critical thinking (pp. 399–401).

The important point is that while the technology can play an important part in facilitating desired learning, the actual learning design and application in practice is likely to be the main arbiter in the actual outcomes. Unfortunately, apart from a lack of good pedagogic design, there has been much criticism of online learning in terms of usability. Shea- Schultz and Fogarty (2002) observed that such basic design failure is common in e-learning environments:

> When most learners complain about e-learning, it's often not the training they object to but the confusing menus, unclear buttons, or illogical links. (p. 117)

Shank and Sitze (2004) emphasize the point that:

> Your success as a designer and developer of online learning is directly tied to your ability to build instructional materials that don't leave users frustrated. (p. 138)

3. The completed blended learning design maximizes the affordances of a range of learning modes and mediums

The recognition that combining modes and mediums of learning may offer more effective (and efficient) approaches to maximizing learning opportunities for a wider range of learners has led to an interest in blended learning designs. As with most things 'educational', there are different conceptions of blended learning. It could be argued that most learning designs are blended in the sense that different methods and resources are typically combined in the creation of teaching and learning strategies. As Littlejohn and Pegler (2007) point out:

> Blending is an art that has been practiced by inspirational teachers for centuries. It centres on the integration of different types of resources and activities within a range of learning environments where learners can interact and build ideas. (p. 1)

In the context of this framework, the following conceptions are acceptable pertinent frames:

> Blended learning is the combination of different training "media" (technologies, activities, and types of events) to create an optimum training for a specific audience (Bersin 2004, xv).

> The goal of blended learning is to synthesize training media into an integrated mix—one you can tailor to create a high impact, efficient and exciting training program. (Bersin 2004, xvi)

The notion of blended learning is attractive, but raises key questions such as:

- What learning outcomes, in which contexts, are best delivered primarily through face-to-face mediums (including blends)?
- What learning outcomes, in which contexts, are best delivered primarily through online platforms (including e-tool blends)?
- How is the complete learning event (e.g., course, module, unit) appropriately structured and managed to maximize the total learning experience for the particular learner cohorts?

Essentially, this boils down to the question of what curriculum components are best delivered in the face-to-face content or in the online environment—the balance of the blend so to speak? From the perspective presented here, the answer is primarily pedagogic and situated rather than numeric. If the following two stages of the design process have been appropriately utilized, this final stage is essentially one of good pedagogic judgement (e.g., application of the core principles of learning), a clear understanding of the affordances of the online learning environment and the specific uses of any particular ICT tool (or set of integrated tools) for enhancing learning effectiveness or efficiency. Furthermore, this will always be in relation to the learning outcomes, the student group, the learning environment and resources access.

However, while this may be an ideal scenario, in practice the "right blend" will depend on many criteria, including the following:

- Programme type and focus (e.g., cost reduction, high impact)
- Learning group (e.g., prior competence, motivational level, cultural factors)
- Resources (including budget and technology infrastructure)
- Content stability (e.g., enduring, relevance).

Within this context of constraints, it is not a question of how much online learning versus how much face-to-face learning; rather about how the face-to-face learning context can be enhanced through ICT and vice versa. Garrison and Kanuka (2004) frame this accurately when they argue that the real indicator of blended learning is not the amount of face-to-face or online learning but their effective integration within a programme.

Invariably, in a specific context, broad decisions can be made about what parts of a curriculum are more likely to be effectively and efficiently deployed online and what is better in the face-to-face medium. For example, course information, syllabi, instructional materials, communication forums and many aspects of the assessment system can often be predominantly online. More interactive learning involving critical discussion and situated interpersonal skills, specialized lab work, role-plays, etc. may be better negotiated in the face-to face context.

In terms of supporting the implementation of CDIO in SP, the use of online workshops is proving useful in a number of ways. Firstly, they provide staff with the key underpinning knowledge for various aspects of CDIO implementation, whether it's for an overview of CDIO or specific areas such as learning design, facilitation skills, performance-based assessment, etc. These have been especially useful for new faculty who were not part of the initial staff development activities. It is intended to have a fully online support facility for all key aspects of implementing CDIO. Secondly, the online workshops also serve as a platform for sharing good practices across schools and departments, as exemplars can be easily uploaded for all faculty to benefit from. The online facility, coupled with specifically tailored short workshops and collaborative project work between subject faculty staff and EDU advisors will evolve into the kind of blended learning model described in this chapter.

5.3 Choosing Technology Tools

In this final section I will offer a frame on certain emerging technologies and their potential use for enhancing teaching and learning. The coverage is not meant to be exhaustive or comprehensive; rather it is deliberatively selective, focusing on the generic principles of ICT affordances, not specific software products. However, having used a range of tools, I will offer my frame on them and what I feel is particularly useful for busy faculty who want the maximum benefits for student learning in the most efficient time-lines possible. You may note at the beginning of this chapter, I made reference to this being the "ideal time for mainstream faculty to really exploit their affordances." Let's see if I can convince you!

5.3.1 Rapid Development Software Tools

In most basic terms these are technologies that enable the production of e-learning content and learning experiences which have the following two key elements:

- Are user-friendly in that only a short time (in some cases only a few hours) is required to master the essential functional use of the technology
- Teaching and learning resources can be produced and up-dated rapidly. This depends on the amount of resources produced, but is significantly quicker than more conventional e-learning development software.

This enables faculty to quickly get up to speed in being able to produce and integrate a variety of media rich and interactive learning resources tailored to course learning outcomes and accommodating to a range of student learning preferences. When used in the context of the design framework outlined in this chapter, they provide real capability for the effective and efficient production, delivery and management of learning experiences.

Apart from the ease of use and rapid production time, they are able to incorporate the key affordances of many ICT documented earlier, for example, "the capability to design multi-sensory learning experiences in the anytime, anyplace, hyperlinked online environment". With a strong pedagogic literacy as the foundational base, teaching faculty can now be really creative in their design of learning experiences for the students they teach.

From my experiences, and remember I am pedagogically focused and not a techie (whatever that means in the present context), apart from PowerPoint which has been around for a long while, I have found the following rapid development software tools to be particularly useful:

SoftChalk LessonBuilder enables the creation of interactive web pages for e-learning courses. The software is easy to use (really) and it enables the quick production of interactive lessons that have a professional look to them. Specific features include pop-up text annotations, self-assessment quizzes, and interactive learning games. After production you can package your lessons for delivery via CD-ROM, Intranet, Internet, or integrate with your LMS (Learning Management System). As their homepage states

> If you can use a word-processing program, you can use LessonBuilder. Designed for teachers and content-experts who don't have time to learn complex software, LessonBuilder is simple, yet powerful, with only the features you need to create exciting, interactive, content for your online course.

They offer free trial downloads and the software is well priced in terms of comparative products on the market. Here's the current homepage: http://www.softchalk.com/.

Camtasia Studio enables the creation of packaged lessons within a self-contained video format that can be web-enabled. It has a screen recording system that will capture a prepared lesson (e.g., Powerpoint presentation) as well as your voice-over during the recording. Using good pedagogic design, high-quality teaching videos can be shared with students on the Web, CD-ROM, as well as on portable media players such as the iPod. The live action video component adds the human touch to the presentation material and enables both technical professionalism as well as the use of informal narrative with humour. The current website is http://www.techsmith.com/camtasia.asp#.

Articulate Presenter is a rapid development software tool that enables quick production of high quality e-learning materials by merging PowerPoint creation with Flash delivery. This enables the quick production of an integrated multimedia presentation that works very effectively due to the format's low bandwidth requirements and the very high (around 98 %) penetration of the cross-platform Flash player.

The natural architecture of the system provides a clear and easy to use menu system and content hierarchy that offers depth of content provision without creating cognitive load and navigational complexity. It offers a very professional and relatively easy to learn rapid development e-learning option, but is comparatively quite expensive. It can be accessed currently at http://www.articulate.com/.

VoiceThread is a web-based application tool that facilitates the presentation of an environment of integrated learning resources (e.g., images, video, documents) in which participants can interact and contribute (e.g., voice, video upload) both synchronously and asynchronously as part of collaborative discussion. It is a more interactive collaboration tool than the other tools outlined. The application is easy to use, provides a versatile learning environment that is easily modifiable and reasonably priced. The current homepage is http://voicethread.com/.

While these applications have some quite different affordances, as indicated in the brief summaries above, all have significant capability to enhance learning effectiveness and efficiency. Given the user-friendly nature of such technologies, it will not be a time consuming or frustrating experience to experiment with them or view some good exemplars from faculty versed in both the technology use and sound pedagogy. From that basis, it should then be readily apparent which applications (and you can use more than one) are best suited for particular parts of your curriculum and for the students you teach.

5.3.2 Social Networking and Collaboration Tools

I don't want to get caught up in definitions here, as Web 2.0 tools cover such a wide range of capabilities and functionality in relation to social networking and collaboration. Precise and uncontested definitions are not possible and indeed, probably not useful. In broad terms Web 2.0 refers to a new, and continually emerging, generation of web development and design tools, that makes possible global communication, information sharing and collaboration on the World Wide Web.

The Web 2.0 tool I find particularly useful is the blog. Firstly, a blog offers the well documented affordances of the online environment identified earlier, such as:

- Anytime, anyplace access to online resources
- Hyperlinked multi-modal, dynamic content
- Global social networking.

This capability in the context of the blog is emphasized by Richardson (2006) when he writes:

> Being able to connect ideas and resources via linking is one of web-logging's most important strengths. (p. 19)

Secondly, the ease of use and familiarity of the blog with young people is highly significant. Blogs are now an everyday part of the communication channels for most students, albeit more of a social rather than educational orientation at present. However, a blog's capability to enhance learning effectiveness and efficiency rates very highly in relation to the core principles of learning. For example, in order to learn a subject effectively, students need to connect new information with what they already know (Core Principle 2: *Learners' prior knowledge is activated and connected to new knowledge*). The very nature and design of a blog facilitates this principle very well. Blog Posts typically appear in reverse chronological order with the most recent post appearing at the top of the web browser. Assuming that a blog was regularly updated each week, students will be able to see a progressive update of the material covered each week and will be able to link what they are currently learning with prior knowledge.

Similarly, as different media types can enhance the learning experience (Core Principle 6: *Instructional methods and presentation mediums engage the range of human senses*, blogs can enable the publishing of a range of multimedia content on the web (video, audio, animation, etc. They can also make explicit and clarify learning outcomes, encourage good thinking, provide rapid, clear and constructive feedback, as well as create a psychological climate which is both success orientated and fun.

Another group of Web 2.0 tools are those focusing on social networking. Facebook is perhaps the most famous example to date. These tools provide a social network service that focuses on building online communities of people who share interests and/or activities. Activities in social networking websites may consist of chats, email, discussion boards, groups, photo sharing, games, etc. As these tools are especially popular with young people, as in the case of the blog, they may be used beyond the more social aspects to a means of communication between faculty and students that have pedagogical benefits also. It may certainly enhance interest and variety to teacher-student engagements.

A final Web 2.0 technology that I will mention in this context is RRS (Really Simple Syndication). RRS enables the pulling together of distributed content from across the web to provide a 'collective intelligence' about certain topics or areas. It can push and pull content, as well as continually update content, from subscribed users.

5.3.3 The Use of the Laptop and Other Mobile Technologies

As laptops have become lighter, more powerful and affordable, especially in the context of widening access and capability of online facilities, so interest in their use has an instructional tool as increased. However, as to yet, there is limited substantive evidence that laptops enhance student learning (Olsen 2002). Furthermore, there is certainly evidence that laptops can be a distracter in the classroom as students may be more interested in personal web-surfing and emailing than the specific lesson experience (Mangan 2001).

While the notebook and other mobile devices do no more than a desktop in terms of capability, they have specific advantages over a fixed workshop station. These are essentially to do with portability and connectivity, which offer learners:

- Convenience
- Expediency
- Immediacy
- Accessibility
- Individuality
- Interactivity.

With increasing portability, the notebook and other mobile devices such as Ipad become an immediate resource platform with the ability to 'push' and 'pull' information for collaborative sharing. Taken together, these affordances create a convenience aspect for the user that becomes a truly personalized computing concept. Both faculty and students can construct their own knowledge bases, share information at will and work collaboratively in the situated context of learning.

However, in that the laptop is essentially no different from any other technology, are there other specific useful pedagogic opportunities that have yet to be recognized or effectively utilized? As I see it, in the present context, apart from the advantages identified above, the main learning affordances of the laptop are in the facilitation of situated learning and the 'here and now' construction of knowledge. For example, while students can of course use the internet out of classroom and on fixed workstations, the opportunity of finding resources and facilitating key understanding at the 'teachable' or 'learnable' moment is one of the most powerful uses of the notebook as an instructional tool. Students can collaboratively find new knowledge and connect to prior learning in the situated context of *the* classroom activity and capture that learning in their own preferred structure (e.g., notes, mind-map, media capture or whatever). Some specific activities include:

- Internet searching for relevant resources in response to authentic learning tasks (e.g., projects, specific tasks, PBL activities)
- Making personal notes, based on classroom discussions and information searched

- Sharing resources in situ with peers and tutor
- Developing media literacy through classroom discussion of different media resources concerning their validity and usefulness.

Invariably, the laptop will become an everyday personal accessory for most students, especially as they are now lighter, more versatile, quicker and perhaps equally important, 'aesthetically stylish.' Similarly, the increasing range of smart mobile devices will offer opportunities for enhancing aspects of the learning process for increasing numbers of learners. However, in the case of mobile learning both common sense and ultimately, research, will show that small screen size will not be that popular in terms of viewing long content or rich media—which is better utilized on desktop or laptop. Even handheld devices that enable students to input or read text in lectures may result in both distractions for lecturers and students. Also, while it may invite some to communicate, it may equally diminish the face-to-face communication—hence its educational value is perhaps questionable, at least in the foreseeable future.

5.4 Summary

This chapter has attempted to provide a practical frame on the use of ICT to support teaching and learning. They can certainly contribute towards learning effectiveness and efficiency in range of delivery modes (e.g., face-to-face, online, blended). However, while the technologies offer many affordances, they are only likely to add significant value to the learning process when employed from a sound pedagogic base.

Most significantly, in terms of the technologies, is the increasing availability of powerful e-learning and web-based applications that are easy to use, providing the 'killer' affordance of creating hyperlinked multi-modal interactive learning environments. While there is no one application that provides all affordances in terms of facilitating learning, it is now a question of selecting and integrating those e-tools that are most useful for the situated content in which you teach (e.g., types of learning outcomes, students taught, and resources capability, etc.).

We now have a readily available suite of tools that can be easily utilized by mainstream teaching faculty as part of their total resource capability as teachers. Apart from the inevitable pressures of time, the capability to produce creative and engaging learning experiences for our students has definitely come of age.

References

Bersin J (2004) The blended learning book. Wiley, San Francisco

Brown T (2009) Change by design. HarperCollins, New York

Clark R, Lyons C (2005) Graphics for learning: proven guidelines for planning, designing, and evaluating visuals in training materials. Pfeiffer, San Francisco

Garrison DR, Kanuka H (2004) Blended learning: uncovering its transformative potential in higher education. Internet High Educ 7(20):95–105
Hamilton S, Zimmerman S (2002) Breaking through zero-sum academics: two students' perspectives on computer-mediated learning environment (Ch 11). In: Rudestam KE, Schoenholtz J (eds) Handbook of online learning: innovations in higher education and corporate training. Sage, London, pp 257–276
Hargreaves A, Fullan M (1998) What's worth fighting for in education. Open University Press, Buckingham
Hodges P, Saba L (2002) Teaching statistics online (Ch 18). In: Rudestam KE, Schoenholtz J (eds) Handbook of online learning: innovations in higher education and corporate training. Sage, London, pp 389–404
Horton W (2006) e-Learning by design. Pfeiffer, San Francisco
Keen A (2007) The cult of the amateur: how today's internet is killing our culture and assaulting our economy. Nicholas Brealey Publishing, London
Littlejohn A, Pegler C (2007) Preparing for blended e-learning. Routeledge, London
Mangan KS (2001) Business schools, fed up with internet use during classes, force students to log off. Chronicle High Educ Sept 7:A43
Marton F (1984) The experience of learning. Scottish Academic Press, Edinburgh
Marzano RJ (1992) A different kind of classroom. ASCD, Alexandria
Melchior T et al (1997) New technologies. In: Costa A, Liebmann RM (eds) Supporting the spirit of learning: when process is content. Corwin Press, California
Oliver R (2007) Describing ICT-based learning designs that promote quality learning outcomes. In: Beetham H, Sharpe R (eds) Rethinking pedagogy for a digital age: designing and delivering e-learning. Routledge, London
Olsen F (2002) Duke U. decides against requiring freshmen to own laptops. Chronicle High Educ Jan 11:A44
Perkins DN (ed) (1995) Software goes to school: teaching for understanding with new technologies. Oxford University Press, Oxford
Richardson W (2006) Blogs, wikis, podcasts, and other powerful web tools for classrooms. Corwin Press, California
Sale D, Cheah SM (2011) Developing critical thinking skills through dynamic simulation using an explicit model of thinking. Paper to be presented at 7th international CDIO conference at DTU Lyngby, Denmark, 20–23 June
Sallis E, Hingley P (1991) College quality assurance systems in mendip paper d20. Coombe Lodge, Bristol
Schrage M (1990) Shared minds: the new technologies of collaboration. Random House, New York
Shank P, Sitze A (2004) Making sense of online learning. Pfeiffer, San Francisco
Shea-Shultz H, Fogarty J (2002) Online learning today. Berret-Koehler, San Francisco
Swartz RJ, Parks S (1994) Infusing critical and creative thinking into content instruction. Critical Thinking Press and Software, Pacific grove
Sylwester R (1995) A celebration of neurons. ASCD, Alexandria
Zemsky P, Massey WF (2004) Thwarted innovation: what happened to e-learning and why. University of Pensylvania, Learning Alliance Report

Chapter 6
Evaluating the CDIO Experience

6.1 Introduction

> Evaluation is the process of delineating, obtaining and providing information useful for making decisions and judgements about educational programmes and curriculum.
>
> (Kemmis 1989, p.117)

Educational initiatives are always to some extent 'experiments'. For this reason it is important to evaluate such initiatives in terms of their impact on student learning. While we may have a well constituted rationale for embarking on an initiative, it is essential to ascertain if the curriculum changes introduced have, in fact, led to the desired improvement in student learning outcomes. Furthermore, we may also want to know more specifically what has improved (or otherwise) and how. From this basis we can better understand how our curriculum and teaching practices are impacting at the level of student experience, and what we might subsequently do to improve them.

The basis for implementing CDIO in the SP context was outlined in Chap. 1. This chapter documents the evaluation approach and outcomes to date; the methodology employed; the main findings and their significance for the future direction of CDIO implementation. The actual evaluation programme, initiated in April 2008, aimed to provide a structured research driven approach to monitor and review the implementation, assessing its worth as an educational framework in the SP context.

At present, the evaluation activities have focused on addressing three broad research questions central to understanding the impact of key aspects of the CDIO implementation:

1. Are the learning outcomes, learning activities and assessments aligned?
2. How have the changes in the curriculum, learning activities and assessments impacted the students?
3. What are the lecturers' perception of the curriculum changes and their impact on students' competence in the selected CDIO skills and interest in the subject?

DOI: 10.1007/978-981-4560-29-0_6,

Table 6.1 Broad and specific research questions

Broad research questions	Specific research questions
1. Are the learning outcomes, learning activities and assessments aligned?	• Are the CDIO skills sufficiently incorporated in the learning outcomes, learning activities and assessments? • Are the learning designs appropriate? • Are the assessments appropriate and valid?
2. How has the changes in the curriculum, learning activities and assessments impacted the students?	• Are the students showing competence in the CDIO skills? • Are the students more engaged and interested? • Do students find the lessons more meaningful?
3. What are the lecturers' perception of the Curriculum changes and their impact on Students' competence in the selected CDIO skills and interest in the subject?	• In what ways, do the activities help develop the selected CDIO skills? • In what ways do the activities encourage interest and learning? • What are the difficulties and areas for Improvement?

The three broad research questions were further refined to more specific questions within each area; these are identified in Table 6.1 below:

6.2 Methodology

The research approach while eclectic emphasized the following key characteristics of qualitative research in that:

- The focus is on the description, understanding and interpretation of human experience in situated contexts (e.g., in this case, student's experience of teachers teaching specific skills)
- Those who are studied are to speak for themselves, to provide their perspectives through personal stories in their own words.

While there are a number of important aspects of curriculum development to be considered, it is the student's perception of the relevance of what was being taught, as well as the actual learning experience in terms of interest, engagement and quality of learning, that is considered paramount in the evaluation. It is one thing to have clearly defined educational aims and objectives, but translating them into effective educational arrangements and practices which actually result in their attainment is, to use an old English metaphor, "another kettle of fish". Furthermore, in the practical context, as Prosser (2008) points out:

> …it is not the way we teach and design our courses that relate to the student learning experience and outcomes, but the way students perceive the teaching and the courses. (p.39)

Different questions in the evaluation required different evidence sources, hence different sampling formats and methods of data collection were incorporated for the various evaluation activities. Where appropriate, triangulation of data has been employed to generate multiple framing and the possibility of enhancing validity in relation to some questions. However, while accessing a range of data sources, we are well aware of Hammersley and Atkinson's (2008) caution that:

> One should not adopt a naively 'optimistic' view that the aggregation of data from different sources will unproblematically add up to produce a more complete picture. (p.199)

Furthermore, while the overall methodology was identified and systematically planned prior to the data collection activities, certain modifications were made in response to emergent data. As Cronbach (1988) points out:

> Designing an evaluation is a continuing process, what variables deserve close attention will be discovered as the fieldwork proceeds. (p.7)

For the qualitative data sources, a broad grounded theory approach (Glasser & Strauss 1967; Strauss and Corbin 1990) was employed in that theory generated will be emergent from the data rather than researcher determined constructs. As Strauss and Corbin (1990) describe:

> A grounded theory is one that is inductively derived from the study of the phenomena it represents. That is, it is discovered, developed, and provisionally verified through systematic data collection and analysis pertaining to that phenomenon. Therefore, data collection, analysis, and theory stand in reciprocal relationship with each other. One does not begin with a theory, then prove it. Rather, one begins with an area of study and what is relevant to that area is allowed to emerge. (p.23)

In the case of the student focus groups, a more phenomenographic approach was adopted after the initial round of interviews. This was in response to recognizing the potential of the research situation to explore more rigorously how students experienced certain key skills of the CDIO curriculum (e.g., thinking) and their perception of how these were being taught. Phenomenography is concerned with describing qualitatively different ways—variation—in which people experience, understand, conceptualize, make sense of various kinds of phenomena in the world around them. While people make sense of the world based on prior experience and selective perception, etc., our common human apparatus and need orientation typically results in shared ways of experiencing the world. As Marton's (1981) suggests:

> phenomena, aspects of reality, are experienced (or conceptualized) in a relatively limited number of qualitatively different ways. (p.181)

This was specifically applied to students' conceptions and approaches to thinking. As critical and creative thinking are an important area of competence in the CDIO syllabus, we were particularly interested in how students experienced the teaching of thinking and their conceptions of 'good thinking'. We sought to identify the variation in the ways in which students went about their thinking and

Table 6.2 Summary of research methods and evidence sources

Broad research questions	Data collection methods (evidence sources)
1. Are the learning outcomes, learning activities and assessments aligned?	• Examination(in collaboration with participating faculty) of a range of curriculum materials(e.g., course documents, module documents, learning plans, schemes of assessment, assessment items)
2. How has the changes in the curriculum, learning activities and assessments impacted the students?	• Student questionnaire for all students in the sample • Student blog • Focus group interviews with a sample of students • Students achievement in assessments(e.g.,performance in Learning activities/test relating to selected CDIO skills)
3. What are the lecturers' perception of the Curriculum changes and their impact on Students' competence in the selected CDIO skills and interest in the subject?	• Focus group interview with faculty teaching on CDIO programmes • Observation of selected lessons(e.g., those incorporating activities related to selected CDIO skills)

how this might inform our practices in teaching good thinking within the subject context.

Table 6.2 summarizes the data collection methods and evidence sources utilized for the various components of the evaluation.

The following sub-sections outline the data collection methods employed, their rationale in context and how they were used in practice.

6.2.1 Examination of a Range of Curriculum Materials

As identified in Table 6.2, this involves the examination of a range of related curriculum materials (e.g., course documents, module documents, learning plans, schemes of assessment, assessment items). The aim was to ensure that the curriculum materials met the conditions of an aligned curriculum (e.g., Biggs 1999) and were consistent with relevant CDIO standards (Crawley et al. 2007).

In practice, this involved on-going collaborative work with school/department faculty (who are the subject specialists) and Educational Development staff (EDU), who provide the pedagogic guidance in producing the necessary materials and to the standard required.

6.2.2 Student Questionnaire

This is a standardized questionnaire administered online via our Blackboard Learning Management System (BLMS) and currently enumerates around 60 % of the full student population (e.g., AY2009/10, 2126 responses, across 17 modules).

The questionnaire employs a number of structured questions, which vary from school/department (e.g., faculty focus and interest) in the form of personalized statements relating to designated experience, which are rated on a 5-point scale (where 5 represents a perception of "strongly agree" and 1 represents "strongly disagree").

It provides quantitative data relating to a number of the research questions across the full student sample. Care is taken in the design of the questionnaire items to ensure clarity, appropriate focus and efficiency, as well as incorporating the wider principles of good questionnaire design and implementation identified in the literature (e.g., Oppenheim 2000). However, it is always useful, in using questionnaires, to be mindful of Quinn Patton's (1983) comment that:

> …questionnaires are probably employed more often than any other technique of evaluation. Questionnaires also probably generate more worthless data than any other technique in evaluation. (p.140)

6.2.3 Student Blog

A student blog, incorporated in the BLMS, was used as a medium for providing ongoing communication and feedback with a sample of student co-participants. Students are typically presented with specific questions relating to their experience of lessons that had selected CDIO skills infused, and asked to provide their responses with examples to illustrate where possible.

They are also at liberty to post comments at any time if they feel this information would enhance our understanding of their learning experience in the classroom context. Apart from the collection of data per se, the use of the blog was seen as a useful and novel way to help build rapport with the students, encouraging more authentic and open communication—hence increasing the possibility of more valid situated data.

6.2.4 Focus Groups

The use of focus groups was employed for the following main reasons:

- Enables the collection of data relatively quickly from a larger number (as compared to individual interviews) of research participants;

Table 6.3 Focus group participants

Semester	Year 1	Year 2	Year 3
1	42	20	4
2	43	38	4

- Provides a more naturalistic context than the individual interview in that it is closer to the everyday conversations that people typically participate in;
- Offers the potential of a synergistic effect in that it allows participants to react to and build upon the responses of other group members, producing richer accounts of the experience being investigated.

The focus group interviews typically lasted from between 1–2 h for both staff and students, depending on situated factors (e.g., the number of participants involved and time commitments, etc.). In practice the process worked well and it was felt that sufficient time is available to encourage a wide range of participation and achieve a sufficient depth of exploration of key areas (e.g., attain "theoretical saturation", Glaser and Strauss 1976; Strauss and Corbin 1990). The general rule in grounded theory research is to sample until theoretical saturation of each category is reached—this means until:

- No new or relevant data seem to emerge, only confirmations of previous data.
- The category development is dense and established.
- The relationship between categories are well established and validated.

The number of students who participated in focused group interviews for the period 2009–2010 is summarized in Table 6.3.

The staff participants in the focus groups comprised those teaching modules in which selected CDIO skills were infused. At the time of writing 28 staff, representing all schools and departments, had participated in the 7 focus group sessions.

In the interview sessions an opening scenario of 3 main areas were presented to staff to offer their experiences and reflections on:

- What have you been involved in doing, in terms of CDIO implementation?
- What have you specifically done and how?
- What is your perception of its impact on student learning based on your experience with student groups?

Where necessary points of clarification are offered, and staff are encouraged to ask their own questions to each other and to participate in the kind of dialogue arrangements that they feel comfortable with. The advice of Douglas (1984), who argues for more "creative interviewing" in which the interviewer must establish a climate for *mutual* disclosure, is interesting in this context:

> Creative interviewing…involves the use of many strategies and tactics of interaction, largely based on an understanding of friendly feelings an intimacy, to optimize cooperative, mutual disclosure and a creative search for mutual understanding. (p.24)

Table 6.4 Student co-participants

Semester	Year 1	Year 2	Year 3
1	153	39	6
2	154	58	6

The interviews are facilitated by two members from EDU; one mainly doing the facilitation and the other recording key responses.

In the case of the student focus group interviews, it was decided at the outset *not* to have students who were 'conscripted'—so to speak—by schools/departments. The ideal situation is one in which the students take on the role of "co-participants" a terms used by Lincoln (1990, p.78), in that they had some personal interest and commitment in taking part in the research activities. They were, therefore, given a full briefing on the research purpose and their role and responsibilities in participating. It was made very clear to the potential student groups that they should only participate if they felt that they could meet the responsibilities in an authentic and conscientious manner. They were specifically required to:

- Chat to classmates and identify their experiences relating to learning the selected CDIO skills and the teaching approaches used.
- Make personal notes and/or blog their experiences in relation to both structured and open questions in the designated student blog.
- Meet with the researchers at least once a semester for group sharing.

The present sample of student co-participants is summarized in Table 6.4 below.

The methodology largely involves questions specifically focusing on their subjective experience relating to aspects of CDIO activities. It is made clear to the students that we are not looking for 'right' or 'better' answers, only the best representation of their experience as they can recall it. The typical interview process for each area of interest involves the following :

- A standard opening scenario is presented and developed with the students (e.g., have you experienced your lecturer explicitly teaching thinking in any of the classes; what does thinking mean to you; how have you responded to this learning experience; what made you experience it in this way, etc.?).
- No new features are introduced—only encouragement for students to further explain and provide examples until no new information is forthcoming.
- Clarification of meaning and checking understanding where appropriate.

6.2.5 Observation of Lessons

The purpose of observing selected lessons taught by faculty involved in the CDIO implementation was to obtain a more ethnographic insight into what is actually occurring in the situated context of the classroom and what might be useful for enhancing understanding of how students experience learning activities related to the selected CDIO skills. Such insights might prove useful in deciding what is of most value in this curriculum and how it is best negotiated in a range of learning contexts. The process for conducting this part of the evaluation is as follows:

- The classes to be observed are mutually agreed by school/department faculty and EDU staff.
- The lesson must incorporate a significant component relating to the teaching/assessment of a CDIO skill area.
- There is opportunity to ask students questions (at the end of the observation) pertaining to their experience of the particular learning activity and CDIO skill areas.

6.3 Data Analysis and Findings

Data analysis techniques were selected on the basis of appropriateness to the data types generated from the various collection methods. Table 6.5 summarizes the approaches taken.

The analysis of the qualitative data components provided the greater challenge in terms of collation and analysis. As Marshall & Rossman (1989) point out:

> Data analysis is the process of bringing order, structure and meaning to the mass of collected data. It is a messy, ambiguous, time consuming, creative, and fascinating process. Qualitative data analysis is a search for general statements about relationships among categories of data; it builds grounded theory. (p.112)

The following sub-sections summarize the main findings from the various data sources obtained in the evaluation to date.

Table 6.5 Summary of data analysis approach

Data type	Data analysis approach
Curriculum materials	Analysis of curriculum documents and materials (e.g., module documents, learning, activities, learning designs, assessment schemes, assessment items and scoring systems) Recording of the numbers of appropriately completed (and non-completed) document/material types
Student blog	Quantitative tabulation and analysis of responses to questions Qualitative data analysis (e.g., categorization and generation of themes)
Student questionnaire	Quantitative tabulation and analysis of responses to questions
Focus groups	Qualitative data analysis (e.g., categorization and generation of themes)
Observation of lessons	Qualitative data analysis using designated recording categories (e.g., tasks relating to thinking, teamwork and communication)

6.3.1 Examination of a Range of Curriculum Materials

This part of the evaluation has been a particularly valuable learning experience, both for faculty and EDU staff. Over the duration of the implementation it became apparent, through the collaborative activity, that a number of modules required significant revision in terms of the writing of learning outcomes generally (e.g., rationalization, performance focus, clarity of intent). There was an initial challenge to both revise and rewrite existing learning outcomes and appropriately infuse selected CDIO skills. The process of rewriting learning outcomes was initially quite time-consuming, both for academic faculty and EDU staff. However, over the past three years many workshops have been conducted on writing learning outcomes and an online tutorial developed, which has reduced the workload significantly.

Once the process of ensuring clarity and appropriateness of learning outcomes (including the infusion of selected CDIO skills) is largely completed, a similar process of review and revision continues for the design of learning tasks and assessment activities (including the scoring systems) until the module curriculum is fully aligned. This collaborative process, apart from the technical curriculum work done, resulted in the building of good rapport between school/department faculty and EDU staff, which proved invaluable over the duration of CDIO implementation.

6.3.2 Student Questionnaire

The following are the summary generic findings:

- Students across schools, understand the usefulness of these skills in their learning and development as technologists.
- The percentage ratings overall tend to be positive (e.g., 4 and 5 combined consistently exceed 3, 2 and 1 combined). However, it is to be noted that there is a significant percentage of 3 ratings across many of the question areas relating to application of the CDIO skills.
- There is some variation in responses across schools and between modules. This is likely to reflect the different approaches taken by schools and departments, as well as the individual faculty effect which became very apparent from the student focus group interviews.

6.3.3 Student Blog

The student blog, in which co-participants respond to their experience in classes with specific CDIO infused skills, proved to be an insightful research component. The student co-participants have shown a high level of diligence in both the regularity and quality of their blog responses. The following are the more generic findings:

- The great majority of students who blogged, perceived the importance of these skills as a valuable part of the curriculum. The variation in perception was low with most students agreeing that these skills are an integral part of being a good engineer as well as useful in a range of life contexts.
- There is considerable variation across modules in terms of student learning experiences relating to these skills. For example, in some modules there is a clear recognition that teamwork and communication are being specifically taught; whereas in others, students feel that they have to just get on with it and sort things out without guidance. It is also evident that there is wide variation, in terms of the student learning experience, emanating from different lecturers' 'teaching styles'. Even within the same class, there is occasionally wide range of perception concerning the usefulness of learning activities. For example in one class, two students made the following comments relating to the same task:

  > I personally feel that it doesn't improve my skill to generate ideas and conceive solutions to engineering problems at all.
  > It has greatly improved my ability…I am able to think out of the box and look into areas for solutions.

- The responses to doing more active and experiential learning (real world tasks) as part of the student learning experience strongly suggests that such tasks link theory to practice, make the learning experience more meaningful and interesting. It is evident that such tasks, when well designed and managed, are effective in both the teaching of CDIO skills as well as consistent with good pedagogic practice generally. One student captured the essence of many when he/she blogged:

  > Making the car chassis helped (sic) learn in groups and communicate well…think critically…improved understanding of how a machine works.

- The explicit teaching of thinking seems to be lacking in many modules. However, a number of responses clearly suggest that some lecturers are systematically encouraging students to think critically through questioning and real world tasks that require good thinking. The following response is not untypical:

  > The lecturers are not exactly teaching us good thinking, but they did a great job making us think by ourselves. The lecturers leave us with questions which we have to crack our brains, brainstorm with our group members and maybe source the internet (sic).

- In response to questions of 'what can be improved', a number of responses point to the importance of the student experience with individual faculty. It is clearly the case that the experience of learning the selected CDIO skills is not separate from the holistic experience that students have of each lecturer. This was very much corroborated from the focus group interviews with students. It is certainly apparent that some faculty are creating a learning experience that is engaging and motivating students (whether these be CDIO related skills or otherwise), while others are not. The latter is aptly captured by one blogger who commented, "Maybe some lecturers should improve their teaching skills".

6.3.4 Student Focus Groups

The student focus groups have been very insightful in terms of understanding the student experience. They provide a means to further unpack the blog response and explore the realities of classroom experience. The following are the more generic summary findings:

- All students who participated in the focus groups felt that the selected CDIO skills (e.g., thinking, communication and teamwork) are relevant and important to learn as part of SP curriculum.
- The experience of learning these skills (indeed, the whole learning experience) seems to be significantly mediated by the particular practices of individual faculty. For example, in some cases, students are clearly experiencing these skills being taught in an explicit manner ("...lecturer poses questions during the practical to probe our thinking. It is on a regular basis"; Mr X challenges us all the time, he wants us to present our thoughts and answers. The way Mr X teaches makes us motivated to learn more"). In contrast, other students depict a less motivating scenario ("some lecturers only talk, talk, talk...give answers without checking for understanding"; "We are told to think well but since we are not guided, we just thinking in the way we want").
- While some faculty are teaching the skills explicitly, this is less evident in others cases. In some situations students are provided with learning activities that involved the skills, but are expected to learn experientially without sufficient guidance and scaffolding.
- When students were specifically probed as to what good thinking involves, it was apparent that only a few faculty had explicitly taught the components of good thinking. The comment by one student, "Lecturer just asks me to think" captured much of the student experience in this area.
- Concerning what constitutes good teamwork and communication, there is again diversity of student experience. Most students who had completed the Teamwork and Communication module felt that it was useful overall. However, in the subject modules, the experience varied from very explicit skill teaching to no explicit teaching at all.

6.3.5 Staff Focus Groups

The response to the questions and issues raised are diverse across schools/ departments and individual faculty. The following represent some of the more salient and general findings across the data:

- The relevance of the CDIO framework (e.g., need to make engineering more practical and interesting) is favourably perceived by staff across the schools/ departments. Some have experienced positive impacts on student attention and interest, especially when doing hands-on activities (e.g., building and racing a car).

The lecturers' experience (e.g., observation and talking to students) supports the view that a greater emphasis on real world engineering projects and activities does result in better student attention and engagement.
- There is agreement that CDIO implementation initially resulted in an increase in workload as a result of the preparation and assessment involved, especially when cohort size is large and there are a number of assessment components. However, one year later approximately half of the staff interviewed felt this had significantly declined as much of the changes that required considerable time (e.g., rewriting module documents, designing assessment) had been completed, and that they were more comfortable with the teaching approaches.
- There is still a range of practices concerning the explicit teaching of the selected CDIO skills. Some faculty are explicitly teaching the skills and following up with relevant performance-based activities. However, in other situations, skills are not being taught explicitly.
- The more motivated students seem to be coping better with the more challenging integrated learning tasks; less competent students are requiring more help, which takes time.
- There is still concern that teaching creativity and other CDIO components may be beyond the existing capability of some faculty. It was reported that some lecturers have insufficient knowledge about effective teamwork and are not comfortable teaching and assessing it. Also, some lecturers find it difficult to develop or find interesting and relevant resources to make learning more active and experiential.
- It is noted that as new faculty come into the course programme they will not have had the training and hands-on experience in teaching CDIO skills. This raises the need for ensuring that appropriate training and support is made available for new faculty.

6.3.6 Observation of Lessons

This evaluation component has been discontinued mainly as a result of time-resource constraints. It was felt that the time invested is unlikely to result in significant new insights relating to the research questions underpinning the evaluation. The main summary findings, based on the sample to date, are:

- Some learning tasks provide opportunities for the development of types of thinking, team-working and communication skills
- The actual skills and what is involved in developing them were not seen to be taught explicitly in most observations. The relative lack of the explicit teaching of these skills is supported from the data obtained from the student focus groups
- Students are generally positive about the learning activities, some needing more support than others.

6.4 Summary and Recommendations

As Kemmis (1989) once argued:

> The quality of the evaluation may be judged by the quality of its contribution to informing and improving the critical debate about the programme. (p.120)

The following summary statements and recommendations are those deemed most pertinent in relation to the research questions, as well as other significant learning derived from the evaluation experience to date.

6.4.1 Research Questions

Q.1 Are the learning outcomes, learning activities and assessments aligned?

The collaborative work between schools and departments in evaluating present curriculum materials (e.g., course documents, module documents, learning activities and assessments) has been an on-going activity over the evaluation period, and is still continuing as new courses join CDIO. As mentioned, this has been and invaluable part of the CDIO implementation, leading to many module documents being improved in terms of alignment as well as establishing good collaboration and rapport between subject teaching faculty and EDU staff. As noted earlier, this was a time-consuming process—albeit an essential part of the implementations success. Increasingly this is becoming less time consuming as school based staff are more familiar with the CDIO framework and are able to take on and coordinate this activity, using EDU more as an overall support function in this area. This should become the norm as a number of school/department staff have worked in close collaboration with EDU staff and we now have a number of good exemplars of practice for reference. For example, many modules, as well as assessment systems and learning activities, are readily available for new staff to review as a basis for reframing their curriculum. Some of these have been identified and illustrated in previous chapters.

Q.2 How has the changes in the curriculum, learning activities and assessments impacted the students?

The data from the range of evaluation activities strongly suggests that there is comprehensive acceptance of the relevance of the infused CDIO skills into the engineering content curriculum. This is consistently articulated in the focus groups and from the blog responses in particular. The quantitative data further supports this across the wider student sample.

In terms of student engagement, interest and perceived meaning, it is clearly the case that real-world projects and other performance-tasks support a positive frame in this context. Most students have communicated that these tasks have made the learning more meaningful, supporting the development of understanding and competence. However, it is also apparent that some students have found the tasks difficult and feel that not enough time is available to fulfil the requirements adequately. This was corroborated through dialogue with staff in their focus groups. It is recommended, therefore, that faculty consider ways to make such tasks more differentiated in terms of requirements (e.g., having certain minimum standards for successful task completion as well as allowing the more competent students to do extension work). This may have implications for staff development in both the design of such tasks, as well as differentiated teaching approaches in their facilitation of projects, etc.

Q.3 What are the lecturers' perception of the curriculum changes and their impact on students' competence in the selected CDIO skills and interest in the subject?

Faculty across schools and departments see the relevance of the underlying purpose and practices of the CDIO framework. There is certainly a high level of agreement that the real world tasks introduced (e.g., building and testing a formula 1 model racing car, etc.) have led to greater student interest and engagement. However, there has been varied experience in terms of actual teaching in the CDIO infused modules. From the earlier evaluation data, workload was a major concern, as staff felt that revising module documents, assessment methods and developing more active and experiential learning designs is time-consuming. While workload concerns still remain, it is less prevalent as an increasing number of staff have completed the document revisions. A significant concern is that some staff report a lack of competence in teaching and assessing certain CDIO skills such as creative thinking and teamwork. While it has been recommended that module teams consider carefully who is best able to teach specific areas of the curriculum, this is not always possible in practice due to manpower considerations and timetabling. It is further recommended that specific training and online support materials are made available to support staff in the most effective and efficient ways possible.

6.5 The Role of the Teacher as the Key Mediator of Experience

A major overall finding from the student co-participants, especially from the focus group interviews, confirms the powerful impact that individual lecturers have on the ways in which students experience their learning in classrooms, irrespective of subject context. While, there is agreement among students that some subjects are "drier" than others, actual faculty teaching significantly impacts attention, subsequent forms of engagement and learning, and the learning approaches adopted.

Many factors invariably influence student learning orientation (e.g., prior experiences, intrinsic motivation), and there is much in the learning situation that can impact what is learnt and how. Most important in the SP context is the quality of teaching, which is consistent with the increasing body of research documented in Chap. 3. The importance of understanding how students are experiencing aspects of our courses and teaching has significant implications for the enhancement of practice and the possibility of helping students to be better learners. As in any human performance activity, there are better and worse ways of doing things—some ways are even dysfunctional to the desired outcomes. The evaluation findings clearly show that there is still some way to go in terms of supporting faculty development in being able to fully utilize the pedagogic approaches and practices outlined in Chap. 3.

While we can never design learning experiences that are going to get desired results with all students, we are now in a more informed position concerning understanding our students' experiences and the meaning they are making of their time in our classrooms. In terms of improving the quality of student learning outcomes, we can now draw upon both a more empirical base of knowledge about human learning, as well as the situated realities of the students we teach, As Marton (2007) suggests:

> It is only when we have a fair understanding of what learners are expected to learn, what they actually learn in those situations and why they learn something in one situation but not in another that pedagogy becomes a reasonably rational set of human activities. (p.22)

6.5.1 The Challenge of Teaching Thinking

It is very apparent, across schools and departments, that many faculty were not explicitly teaching the CDIO skills infused. This is evidenced most strongly from the student focus groups and verified by some staff. While many of the modules incorporate learning activities that involve the skills, in many cases it is assumed that students will learn them implicitly. From the student feedback, this is very often not the case, particularly noticeable in relation to thinking. When asked, the majority of students interviewed still had limited and idiosyncratic perceptions relating to what is good thinking.

The student response to the question, what is 'good thinking', suggests significant variation in constructs relating to what thinking is and its underlying cognitive heuristics. The following examples of student response illustrate this variation:

> conscious mind, a good amount of reasoning
> using the mind like a freight train with a checklist
> logically, systematically and creatively to solve problems
> thinking in positive and optimistic ways

Students' prior learning (including the impact of the explicit teaching of thinking) plays a significant role in their present framing of what is good thinking. For example, from the wider evaluation evidence, it is clearly the case that some staff had explicitly taught the components of good thinking as identified in Chap. 2 (e.g., specific types of thinking, dispositions relating to good thinking). It may well be that this is resulting in some students applying the model and incorporating the 'language of thinking' into their descriptions. In contrast, many staff had not taught thinking explicitly, which may equally be related to the less formalized constructs of what is good thinking as depicted by many students.

6.6 Summary Frame on Future Directions

While it is important that the evaluation stays focused on the key research areas already identified, it is necessary to both move away from questions that have become 'theoretically saturated'—so to speak, for exploring other emerging questions relevant to a better understanding of the specifics of the students learning experiences. For example, we have decided to discontinue the present evaluation approach with first year students as it is now apparent that the likelihood of new insightful data emerging is unlikely, and subsequent time spent would not be cost-effective. We have made the inference and interpretation that the majority of students perceive the relevance of CDIO skills and the curriculum approach employed to make their learning more interesting and real world focused. At present, we are focusing the evaluation more on the student's actual experience while doing specific types of learning activities (e.g., projects, cases) to attain better insight into factors that may enhance their level of intrinsic motivation.

This data from the evaluation programme so far supports the usefulness of certain pedagogic and assessment practices relating to CDIO standards. For example, well constituted integrated real world tasks, when effectively managed, typically result in better student interest and engagement. There is also evidence, based on faculty reporting, that students are displaying better learning outcomes in terms of understanding and actual competence.

However, the evaluation has revealed that there are systemic issues relating to teaching and learning, which may require more focused and creative solutions. Based on our existing framing, the following remain pertinent questions and issues to pursue over the next evaluation cycle:

- Are students developing a real understanding of the selected CDIO skills and able to display both competence and transfer, and to what extent? While there is evidence that this is occurring in some classes, it is clearly mediated by the pedagogical and related communication styles of faculty. It is important to further explore what types of learning arrangements and pedagogic competences are most conducive to better learning. The approach documented in Chap. 3 is particularly pertinent in this context.
- A consistent finding is that students' perception of what is good thinking shows a high degree of variation. It is clearly evident that while the development of good thinking (i.e., critical, creative, metacognitive) remains an educational aim, it is not established as consistent pedagogic practice at present. The challenge of contextualizing good thinking to the various subject field contexts

will be an interesting and necessary part of forthcoming professional development activity.

- How to develop the necessary staff competence to meet the demands and rigor of the increasingly challenging professional teaching role? What are the most effective and efficient structural arrangements, learning platforms and modes of delivery, etc.?

References

Biggs J (1999) Teaching for quality learning at university. Open University Press, Buckingham

Crawley E, Malmqvist J, Ostlund S, Brodeur D (2007) Rethinking engineering education. Springer, New York

Cronbach L (1988) 'Issues in planning evaluation'. In: Murphy R and Torrance H Evaluating education: issues and methods. Paul Chapman Publishing, London

Douglas J (1984) Creative interviewing. Sage, London

Glaser B, Strauss A (1967) The discovery of grounded theory. Aldine, Chicago

Hammersley M, Atkinson P (2008) Ethnography: principles in practice. In: Silverman D (ed) Doing qualitative research: a practical guide. Sage, London

Kemmis S (1989) 'Seven principles for programme evaluation in curriculum development and innovation'. In: House ER New directions in educational evaluation. The Falmer Press, London

Lincoln YS (1990) The making of a constructivist: a remembrance of transformations past. In: Guba EG (ed) The paradigm dialog. Sage, London

Marshall C, Rossman GB (1989) Designing qualitative research. Sage, London

Marton F (1981) 'Phenomenography: Describing conceptions of the world around us'. In: Instructional science 10, 177–200. Elsevier Scientific Publishing Company, Amsterdam

Marton F (2007) Towards a pedagogic theory of learning. In: Entwistle N, Tomlinson P (eds) Student learning and university teaching. Leicester, The British Psychological Society

Oppenheim AN (2000) Questionnaire design. International Publishing Group, London

Prosser M (2008) 'Experiencing teaching and learning in higher education'. In: Frontiers in High Educ. International Conference on Teaching and Learning in Higher Education, CDTL, Singapore

Quinn Patton M (1983) Practical evaluation. Sage, London

Strauss A, Corbin J (1990) Basics of qualitative research. Sage, London

Chapter 7
Managaing the Change Process; Approach, Strategies and Professional Development

7.1 Introduction

> ...where educational change is concerned, if a teacher can't or won't do it, it simply can't be done. (Hargreaves and Evans 1997, p. 3)

The importance of effective management of the change process, as well as the various technical aspects of the change itself, is well documented in the literature (Fullan 2007; Stacey 1996; Hargreaves 1994). Equally, the establishment of a professional development approach to support emerging staff learning needs, in ways that are effective and practical from the faculty perspective, is also of key importance in the change stakes. In this chapter, I will outline the approach taken, the key strategies employed and the professional development framework that has evolved to support the implementation of CDIO.

In any change initiative, there are certain general questions that need to be thoughtfully addressed at the outset, these include:

- Identifying what needs to be done, by who and how it might be done—possible approaches, strategies, etc
- Anticipating the possible consequences of the different approaches that can be taken and the strategies adopted
- Choosing the preferred approach, strategies, etc.

In practice, while educational innovations are inevitably multidimensional, they typically encompass at least three components:

- New or revised curriculum and materials
- New teaching approaches
- Alteration of beliefs (e.g., pedagogical assumptions about teaching and learning).

As CDIO implementation involves significant curriculum restructuring, addressing these three components is central to the change process. In working towards meeting the CDIO Standards, much work has gone into customizing the general syllabus into one that was relevant and practical to the polytechnic context and which could be further contextualized to the various engineering schools'

D. Sale, *The Challenge of Reframing Engineering Education*,
DOI: 10.1007/978-981-4560-29-0_7,

context. As documented in Chap. 2, this was time-consuming, but worth the effort spent in terms of developing faculty ownership, collaborative empowerment and, not least, a belief that the curriculum change was achievable and useful.

Similarly, for some faculty, CDIO implementation created challenges to aspects of their traditional teaching practices, as well as underlying beliefs about teaching and what this entails. Having to adopt a more holistic educational approach was, and still is for some, rather vague and questioned in terms of viability in the polytechnic context. It became very apparent from the outset that, apart from dealing with any technical aspects of enhancing teaching knowledge, pedagogic skills and techniques, a bigger challenge would present itself at the level of faculty perception and beliefs. Sustained and lasting change needs to occur across all components for it to be effective in practice. For example, there is little point in successfully changing the curriculum documents and materials without corresponding changes in teaching approaches, practices and accompanying pedagogical beliefs. Most fundamentally, as Fullan (2007) has argued:

> …changes in beliefs and understanding (first principles) are the foundation of achieving lasting reform. (p. 33)

In broad terms, the change was framed in terms of successfully navigating these three main components through the change processes of initiation, implementation and institutionalization (Fig. 7.1: Summary of the Change Process) that have been well documented by Fullan (2007).

Change can also be usefully conceptualized in terms of different levels of human engagement. For example, Hersey and Blanchard's (2001) depiction of four main levels of change (Fig. 7.2) illustrates that while each level is important and must be effectively negotiated in the change management strategy, progression down the levels is paramount for meaningful sustained change.

Change at the 'knowledge' level refers to stakeholder awareness of the change, its goal, purposes and key activities involved. At the 'attitudes' level, stakeholders have developed a personal response to the change in terms of how they feel about it, which needs to be generally positive if the change is to proceed as planned. Change at the 'individual behaviour' level refers to some stakeholders effectively implementing the practices underpinning the change. Changes at 'group/organizational behaviour' level can be considered to have resulted only when a critical mass of stakeholders (i.e., enough to make a significant impact in practice—however defined) are behaving in accordance with the change objectives.

Invariably change becomes more difficult as we move down the hierarchy. Also, without significant change in behaviour at the group/organization level, the initiative is unlikely to reach a significant level of sustainability. Furthermore, even if the desired change has become well established in terms of organizational behaviour, there is always the issue of maintenance. Successful institutionalization of a change initiative does not guarantee further progression or even continuity in the same way that reaching a high level of physical fitness does not ensure subsequent fitness. Huberman and Miles (1984) identified a number of factors that can

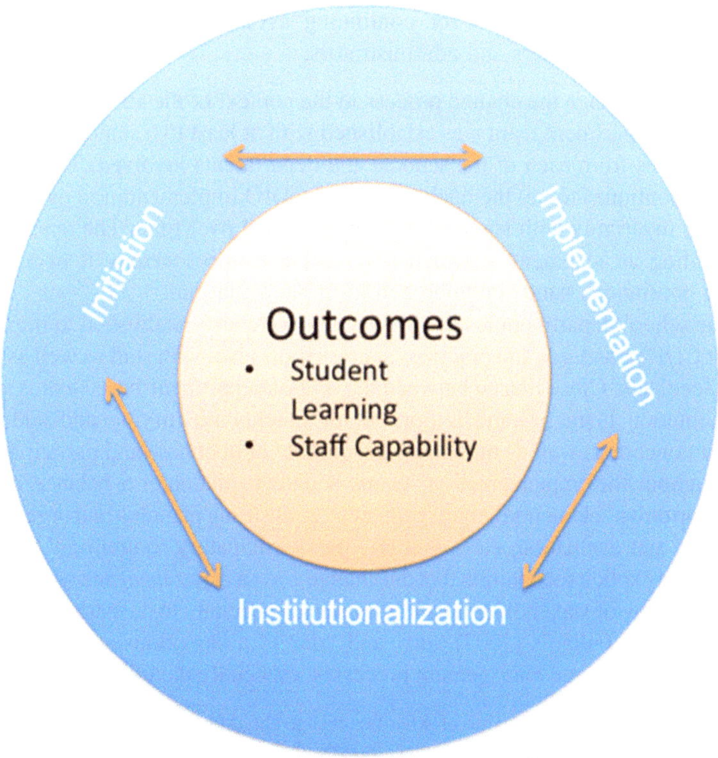

Fig. 7.1 Summary of the Change Process

Fig. 7.2 Hersey and Blanchard's four main levels of change

contribute significantly to the continued success (or otherwise) of an innovation over time. Much depends on whether or not the change:

- Gets embedded or built into the structure (through policy, budget, timetable, etc.)
- Generates a critical mass of administrators and teachers skilled and committed to the change

- Has established procedures for continuing assistance, especially relative to supporting new teachers and administrators.

In order to manage the change process in the context of the above framing, a SP inter-school/department team was established with at least two senior management representatives from each of the schools and departments involved. This team has been (and continues to be) the anchor point of CDIO implementation over the past five years, invariably with some change in personnel over time. The team secured a timetabling arrangement in which it was able to meet weekly if necessary to ensure a continuous hands-on approach to implementation.

At the school/department level, CDIO teams were also established to implement relevant CDIO standards and practices at grass roots level with staff as well as obtain faculty feedback. Close liaison between the school/department-based teams and the implementation team became important for quickly identifying and addressing issues of concern as well as maintaining a climate of openness and participation.

Throughout the implementation, teams sought to maintain a balance between certain 'top down' elements of management (e.g., ensuring planning and organization, monitoring and evaluation, etc.) with an open participatory 'bottom up' approach. There was an explicit recognition that the change was an *experiment* and would require much in the way of stakeholder meaning-making, especially in the early stages of the implementation. Fullan's (2007) "do" and "don't" assumptions are worth careful reflection in the change management processes, especially the following:

> Do not assume that your version of what the change should be is the one that should or could be implemented. On the contrary, assume that one of the main purposes of the implementation is to *exchange your reality* of what should be with the realities of implementers and others concerned, through interaction with them.
>
> Assume that people need pressure to change (even in directions that they desire), but it will be effective only under conditions that allow them to react, to form their own position, to interact with other implementers, to obtain assistance, to develop new capabilities…
>
> Assume that effective change takes time. It is a process of "development in use." Unrealistic or undefined time lines fail to recognize that implementation occurs developmentally. Significant change in the form of implementing specific innovations can be expected to take 2 or 3 years;… (p. 123)

It is important, therefore, to focus on what needs to be done properly, at least from the particular framing at the time, rather than rush things through to give an impression that lots of activity is occurring. We tried to keep to the maxim that 'more is *not* better, better is better.' It's a challenging one in practice, believe me.

7.2 Managing the Change Process

In any change initiative there are always questions of why is the change necessary and for what purpose. As mentioned in a previous chapter, education is often 'a creature of fashion' and folk who have been in the profession for some time are often suspicious of large scale curriculum innovations, especially if presented as "the best thing since sliced". Hence, questions of why CDIO, how can it improve aspects of professional practice and the student learning experience are central at the initiation stage. Indeed, they periodically re-emerge as new faculty become involved in CDIO implementation.

There was a clear recognition that simply spelling out the need and rationale of CDIO, as documented in the pioneering work of Crawley et al. (2007), despite being grounded in the engineering context, would not be unproblematic. As a result, we avoided over-hyping CDIO, focusing on what we felt were the pertinent issues for wider faculty discussion and engagement. Firstly, as CDIO is an educational framework conceived by engineers for engineering education, it has greater face-validity than more generalized curriculum change initiatives. Secondly, the framework being interpretive rather than prescriptive, and explicitly encouraging customization to the local context added to the likelihood of acceptance.

Furthermore, while CDIO had been identified as a possible framework for curriculum improvement, it was not mandated by senior management, rather encouraged on a volunteer basis. Schools/departments could choose their level of participation and select courses for pilot work. This made possible an initiation approach that encouraged open discourse on the various aspects and rationale, and how it might be useful in the SP context.

This approach proved significant in terms of reducing the level of resistance, especially over time. In retrospect, we were able to openly and collaboratively consider the adoption of CDIO in terms of its use as a means of improving the educational opportunities of our students. We certainly could not be accused of, to use a Shakespearian quote "me think he doth protest too much" in terms of CDIO advocacy. While faculty varied considerably in terms of their active involvement in developing areas of practice consistent with the standards, there has been little overt resistance to the goals and approach of CDIO. At present, course teams are asking to be involved in CDIO as they perceive benefits in terms of student learning, especially in the context of the evaluation results over the past 3 years.

Other well documented characteristics of change such as clarity and complexity were also relevant considerations to the implementation of CDIO. Questions of what needs to be done and how have proved to be challenging. For example, while certain 'CDIO Skills' were identified as most useful and pertinent for infusion into course curricula, as documented in Chap. 2, some faculty remained unsure of what this actually meant in terms of course and module restructuring, learning design, actual teaching and assessment. Ensuring clarity and minimizing complexity are crucial, but unfortunately not so easy in practice. As Charles Mingis (2001) once stated:

> Making the simple complicated is commonplace; making the complicated simple, awesomely simple, that's creativity.

In order to conceptually frame the most significant factors that might support the implementation, as well as possible resistors, a force-field analysis was periodically conducted to identify existing perceptions of the present change scenario in relation to the desired situation. Force field analysis involves the use of a graphical organizer (tool) to frame the various forces (as perceived at any point in time), identifying those forces that are seen to support the change and those that may be resistors to the change. The idea, in basic terms, is to maximize the leverage on the forces supporting the change and mitigate or 'turn round' (make into positives) those that may constitute barriers. Figure 7.3 shows the basic force-field analysis template:

The initial use of the tool, as captured in the summary diagram (Fig. 7.4), identified the following salient forces:

The forces supporting the change have remained strong in terms of perceived potency. While there are still some faculty who share the view that SP has been successful for over the past half-century, and there is no need for change, most now recognize that change is necessary to remain relevant and competitive in the educational provision stakes. Furthermore, a greater awareness that lecturing in

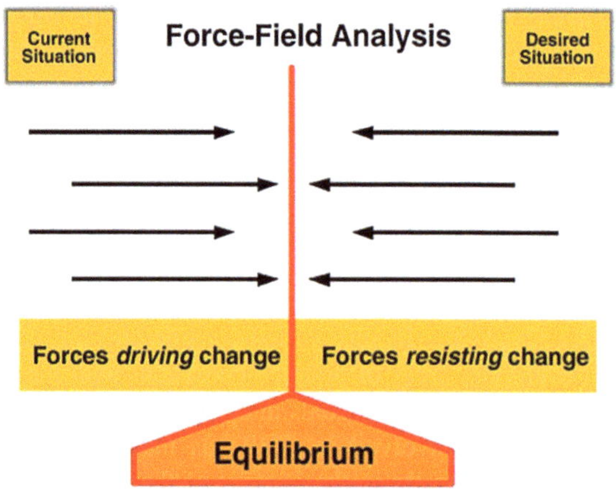

Fig. 7.3 Force-field analysis Template

Forces Supporting the Change:	Forces Resisting the Change:
• Desire for a more relevant (21st century skills based) engineering curriculum (4)	• Modularization and silo mentality (4)
• Promotion of creativity, innovation and enterprise in Singapore (4)	• Perception that teaching should focus on subject technical content and skills (4)
• CDIO Champions (3)	• Heavy staff workloads (4)
• Making engineering more interesting to students (3)	• Professional development in pedagogy not seen as important as field specific professional development (3)
• Involvement in CDIO community (3)	

Fig. 7.4 Summary of forces supporting and resisting the change

higher education requires much more than transmitting technical content, but actually teaching in the more holistic sense, has become less of a resistor and potentially an enabler. For example, SP's "Towards SP@60 (2014)", strategic planning document (2009/2010) identified 'Holistic Education' as a main thrust for the polytechnic. There is now a greater focus on making teaching more active and experiential as well as being more responsive to the social and emotional aspects of student learning. This in turn, brings into focus a heightened awareness of pedagogic knowledge and its importance in the design of learning experiences.

Over the past 3 years, significant progress has been made in terms of curriculum integration and developing integrated learning experiences, most significantly through design-implement and multi-disciplinary projects. Staff workload has remained a significant issue of concern (as depicted in the previous chapter). The establishment of Academic Mentors in schools/departments, who have responsibility for charting educational initiatives at school level may, in the longer term, result in a heightened perception by faculty of the importance of continual professional development in pedagogy, along with technical up-grading.

7.3 Meaning Making and Relationship Formation

Much of the literature on organizational behaviour paints a daunting picture for those expecting, even desired, change to be a smooth process (e.g. Brunsson 2000; Stacey 1996; Argyris 1993). Stacey (1996), from a complexity theory standpoint, argues that paradoxes inherent in the human condition make organizational life far from one of stability and equilibrium. He identifies these as:

- The capacity to both love and hate at the same time
- The urge to separate from others and individuate and the longing to fuse and lose oneself
- The capacity for destructiveness and the creative urge to make reparation
- The need for safety and comfort of the known and the desire for the exciting and the unknown
- The desire for the rational, the functional, and the factual, on the one hand, and the longing for the aesthetic, the emotional, and the sacred on the other (p. 130).

He goes on to argue that:

> …through this lens, the world of organizations is seen as a system held far from equilibrium, at the edge of chaos, by the paradoxical dynamics of competition and self-organizing cooperation. In this fundamentally paradoxical world, the links between actions and their long-term outcomes are lost in the complex interactions between various components of the system. (p. 248)

In a similar vein, Argyris (1973) goes as far as asserting that:

> ...trust, openness, and risk taking are rare in formal organizations, and they are significantly deviant from the mistrust, closedness, and emphasis on stability which we suggest is more typical. (p. 82)

However, in practice, one has to work with reality as it presents itself, not how one might want it to be. Certainly an understanding of how human behaviour actually plays out in organizational contexts at least mitigates surprise and disappointment when things appear chaotic. In recognition of this, we sought to work carefully through the curriculum restructuring stages, negotiating key processes and identifying professional development activities as need emerged. The key phases of the curriculum reorganization are summarized below:

- Awareness of the CDIO educational framework, what it was, how it works and in what ways might it support SP curriculum goals
- Reframing the curriculum in terms of learning outcomes, structure and sequence
- Designing the learning experience, incorporating methods, activities, and resources
- Production of aligned assessment systems, methods and instruments
- On-going evaluation, incorporating the student experience, faculty experience, curriculum materials and student performance.

A strategic aspect that proved important in effective implementation was the early involvement of faculty in a manageable part of the change activity. Once induction had reached a level where most faculty were aware of what CDIO was, its potential benefits and the key phases of curriculum reorganization, etc., there was a push for them to quickly get involved in useful CDIO related activity. In reviewing courses and modules, for example, they were able to experience tangible results in terms of clearer and more meaningful documents, from which to plan student learning. As the old saying goes, "The proof of the pudding is in the eating" or as Bate et al. (2005) illustrate in more technical terms:

> ...people cannot want it until they have tried it. The concrete experience of participating in a movement is crucial, meanings and value being formed after the experience not before it. (p. 31)

Similarly, Guskey (2000) makes the point that educators do not typically change their beliefs from most professional development opportunities. Their practice is only likely to change when they see evidence that the change positively affects student learning. This process of practical engagement within realistic time-frames was continued throughout the key implementation phases. Evidence of successful implementation could be experienced in terms of curriculum products, and direct experience in classroom practice and student response. There are, as Fullan (2007) rightly points out, "no shortcuts to achieving shared meaning, including serving it on a platter" (p. 104).

A second key strategy has been a focus on developing and maintaining good relationships with all stakeholders. Over the past five years, the implementation team has worked collaboratively to make meaning of the CDIO experience and try to ensure that action taken is realistic and practical both in terms of school/

department reference norms and for SP generally. This inevitably involved much discussion, disagreement and negotiation. However, a significant factor in terms of success has been the nature of relationships between the various team members in terms of positive interpersonal interactions. Despite the number of meetings, constant revisiting of what we were doing and the inevitable tedium that is sometimes the reality of meetings, there was acceptance of difference, management of egos (even mine) and, not infrequently, humour. The ability to "see the funny side" is important in work and life, and especially in managing change.

At a personal level, good rapport with a number of colleagues involved in the implementation, both with the implementation team and school/department-based faculty has resulted in several papers co-written with colleagues. I had not previously written papers with engineers about engineering education. Hopefully, they may have learned some useful knowledge about human learning and pedagogy from me, as I certainly know much more about engineering. The observation by Hargreaves and Fullan (1998) appears to be well constituted:

> Structures are only as good as the relationships and know-how of the people who occupy them. Emotional management is ultimately about attending to these relationships properly. Managing emotionally and rationally in today's turbulent times is rocket science. (p. 129)

Similarly, Robbins (2001) powerful observation is worth reiterating in this context:

> Rapport is the ultimate tool for producing results with other people. (p. 231)

7.4 Professional Development

There are many 'resistors' to change, as well as the reasons for resistance. Resistance to change is typically increased when people:

- Don't see the purpose of the change
- Feel uninvolved in the change process
- Lack skills that are necessary to bring the change about
- Are unclear about what needs to be changed and how to do it.

While lacking the necessary skills and clarity of what to do and how are typically perceived as pertinent issues for professional development, I take the position that all aspects of resistance are professional development concerns. If professional development is to have practical value it must result in contributing to the core business of the organization. In educational institutions, that core business is the quality of student learning outcomes (however defined). Invariably, there is much that contributes to this overarching goal, not least the quality of teaching and related curriculum components. Professional development, while necessitating customization to the particular needs and role context of individual faculty, is ultimately a collective and shared capability. It must encompass the engagement of key stakeholders in the very process of making meaning and deciding the worth of a change initiative which, in turn, involves the critical review of educational goals, practices, research into the proposed change, and some good thinking.

The professional development approach deemed most useful initially was to be quickly responsive to faculty needs and concerns with the provision of a range of staff development functions (whether these be customized workshops, working with school faculty on the development of curriculum materials, or other resource provision). Most significant, in terms of ensuring success, is the on-going collaborative work between implementing faculty and the education advisors. As the education advisors had been involved with both school and department representatives from the outset it became the practice to further develop and leverage on this close liaison throughout the implementation. There was a clear recognition that professional development activities should evolve based on faculty perception of what is needed and most useful at particular phases of the implementation. For example, in the initial phase of infusing the CDIO skills in course and module documents, much collaborative work between school faculty and education advisors was done, and subsequently followed by sharing and evaluating examples of practice. The same process continued for the production of learning activities and assessments.

At the present time, school/department based faculty continue to work collaboratively with EDU education advisors to support all aspects of CDIO implementation, whether it is induction of new faculty or further development of mature programmes. Our approach to professional development is in many ways akin to that summarized by Elmore and Burney (1999):

> We know a good deal about the characteristics of successful professional development: it focuses on concrete classroom applications of general ideas;it exposes teachers to actual practice rather than descriptions of practice;it offers opportunity for observation, critique and reflection; it providesopportunity for group support and collaboration; and it involves deliberate evaluation and feedback by skilled practitioners with expertise aboutgood thinking. (p. 263)

A similar frame is suggested by Darling-Hammond and Bransford (2005) who summarize that:

> Emerging evidence suggests that teachers benefit from participating in theculture of teaching–by working with the *materials and tools of teaching practice*;examining teaching plans and student learning while immersed in theory aboutlearning, development and subject matter. They also benefit from *participatingin practice* as they observe teaching, work closely with experienced teachers, and work with students to use what they are learning. (p. 404)

In implementing CDIO, there has emerged an essential need to be able to rise to the challenges of an increasingly diverse and challenging professional teaching role. While the traditional lecturing approach, that has been the norm for many faculty, still has merit as an important teaching method, it is no longer a sufficient pedagogy for developing the kind of student competency profile needed for the modern workplace and life preparation generally. Furthermore, as discussed in detail in Chaps. 3 and 5, we are now in a better position to provide more effective and engaging learning experiences through good pedagogic design and the affordances of information-communication technologies. The implementation of CDIO has coincided with an increasing refocusing of educational quality towards that of enhancing the pedagogic competence of teaching professionals. The importance of teaching quality has been clearly established by Izumi and Evers (2002) who, drawing from an extensive range of research source, concluded that:

> What the research does show is that the quality of classroom teachers has the greatest impact on the performance levels of students. (p. xiii)

7.5 Summary

This chapter has attempted to provide a practical frame on managing the change process in implementing CDIO, as well as our response to ensuring appropriate and viable professional development for the faculty involved. Key aspects of the change processes have been highlighted as well as core strategies employed. The case is made for a thoughtful approach that involves both systematic planning and a preparedness to respond quickly and responsibly to the emerging experiences, concerns and needs of implementing faculty. The importance of allowing time and providing on-going support for faculty to make meaning of the change objectives, understand what needs to be done and how, as well as develop the necessary skills for effective implementation is paramount. We still have some way to go in terms of the change becoming institutionalized, let alone sustained. However, we feel that much of the implementation approach has been conducted using a sound curriculum development framework, pedagogic understanding and, most significantly, good principles of human conduct. Perhaps that's as good as it gets in the real world of educational practice, or in any professional context come to that. We intend to continue in the same vein.

References

Argyris C (1973) Intervention theory and method. Addisson-Wesley Publishing Company, London

Argyris C (1993) On organizational learning. Blackwell Publishers, Cambridge

Bate P, Bevan H, Robert G (2005) Toward million change agents: a review of the social movements literature. National Health System, London

Brunsson N (2000) The irrational organization: irrationality as a basis of organizational action and change. Copenhagen Business School Press, Copenhagen

Crawley E, Malmqvist J, Ostlund S, Brodeur D (2007) Rethinking engineering education. Springer, New York

Darling-Hammond L, Bransford J (2005) Preparing teachers for a changing world. What teachers should learn and be able to do. San-Francisco : Jossey-Bass

Elmore RF, Burney D (1999) Investing in Teacher Learning. In: Darling-Hammond L, Sykes G (eds) Teaching as the learning profession. Jossey Bass, San Francisco
Fullan M (2007) The NEW meaning of educational change. Teachers College Press, New York
Guskey TR (2000) Evaluating professional development. Corwin Press, Thousand Oaks
Hargreaves A, Fullan M (1998) What's worth fighting for in education. Open University Press, Buckingham
Hargreaves A, Evans R (1997) Beyond educational reform: bringing teachers back in. Open University Press, Buckingham
Hargeaves A (1994) Changing teachers, changing times: teachers' work and culture in the postmodern age. Teachers College Press, New York
Hersey P, Blanchard KH (2001) Management of Organizational Behaviour. Prentice-Hall, Englewood Cliffs
Huberman M, Miles M (1984) Innovation up close. Plenum, New York
Izumi TL, Evers WM (2002) Teaching quality. Hoover International Press, San Francisco
Mingis C (2001) Quote. http://www.brainyquote.com/quotes/authors/c/charles_mingus.html Accessed 18 Aug 2010
Robbins A (2001) Unlimited power. Pocket books, London
Stacey R (1996) Complexity and creativity in organizations. Berrett-Koehler, San Francisco
Towards SP@60(2014), (2009/2010), *Opening minds, shaping lives: the strategic plan of singapore polytechnic*. Singapore Polytechnic, Singapore

Appendix 1
The CDIO Syllabus Customized for Singapore Polytechnic

Technical Knowledge and Reasoning

Knowledge of Underlying Sciences [a]

Mathematics (including statistics)
Physics
Chemistry
Biology.

Core Engineering Fundamental Knowledge [a]

Fluid Mechanics
Solid Mechanics and Materials
Dynamics
Signals and Systems
Thermodynamics
Control
Computers and computation.

Advanced Engineering Fundamental Knowledge [k]

Aerodynamics
Structural Mechanics
Structures and Materials
Jet and Rocket Propulsion
Flight and Advanced Aerospace Dynamics
Computational Techniques
Estimation and Navigation
Human and Supervisory Control
Digital Communications

D. Sale, *The Challenge of Reframing Engineering Education*,
DOI: 10.1007/978-981-4560-29-0,

Software Engineering
Autonomy
Digital Circuits and Systems.

Personal and Professional Skills and Attributes

Engineering Reasoning and Problem Solving Process

Identify and formulate problems

Evaluate background data and symptoms
Select the key issues to be resolved
Write a problem statement.

Formulate a strategy to solve problems

Apply tests for consistency and errors (limits, units, etc.)
Generate alternative solutions (using models, analytical and numerical solutions, qualitative analysis, experimentation and consideration of uncertainty, working within realistic time constraints)
Analyse essential results of solutions and tests
Analyse and reconcile discrepancies in results
Formulate summary recommendations.

Experimentation and Knowledge Discovery

Formulate hypothesis

Select critical questions to be examined
State hypotheses to be tested.

Conduct Literature Review

Conduct information search and identification using library tools (on-line catalogs, databases, search engines)
Sort and classify secondary information
Evaluate the validity and reliability of information
Identify essential information (e.g., innovations) contained in the information
Cite relevant sources of information.

Conduct Experimental Inquiry

Construct experimental design (e.g., Identify controls and control groups
Identify valid sources of data, Identify the precautions (e.g., environmental, safety and health) in conducting experiments

Conduct experiment (Apply test protocols and experimental procedures, Collect and collate data).

Analyze data and write report

Validate data using statistical methods
Identify the limitations of data employed
Draw conclusions supported by data
Appraise possible improvements in knowledge discovery process.

System Thinking

Understand the basis and methods of System Thinking

Explain the purpose, discipline and methodology of system thinking
Identify types of systems, sub-systems and components
Identify the key characteristics ('laws') of systems
Apply systems thinking in problem-solving.

Use a range of systems thinking tools

Identify factors that affect the behavior of the system (e.g., "input", "throughput", "output", "feedback", and "delay")
Analyse the impact of feedback and delay on system behavior.

Personal Skills and Attitudes

Apply the thinking process

Identify key thinking skills in good thinking
Identify barriers to effective thinking (e.g., traits, dispositions, working memory, perception, lack of information, etc.)
Identify factors that promote effective thinking (motivation, openness, risk taking, exposure to varied knowledge bases and ideas, etc.)
Use a range of critical thinking skills (e.g., analysis, comparison and contrast, inference and interpretation, and evaluation)
Use the creative thinking process (e.g., generating possibilities, incubation, illumination)
Use a range of creative thinking tools and techniques (e.g., Brainstorming, Mindmapping, TRIZ)
Identify contradictory perspectives and underlying assumptions
Reframe and take a range of different perspectives
Use metacognition in monitoring the quality of personal thinking.

Manage Learning

Identify key aspects of the learning process

Identify how emotions and beliefs affect learning
Identify approaches for self-improvement (e.g., lifelong learning, creating positive beliefs and psychological states, managing emotions, etc.)
Display key dispositions (e.g., initiative, perseverance, flexibility, etc.) in work projects
Use a range of learning strategies and skills (e.g., goal setting, learning plans, organizing/summarizing information, receiving feedback, etc.)
Manage time and resources.

Professional Skills and Attitudes

Evaluate the impact of values and ethics

Identify the need for values and ethical codes of conduct
Compare and contrast value systems and ethical codes of conduct
Analyse the impact of values and ethical codes of conduct on personal and professional behaviour.

Demonstrate professional behavior at work and in society

Use ethical reasoning on issues relating to human conduct in personal and professional contexts
Identify behaviours that demonstrate social responsibility
Demonstrate behavior consistent with agreed codes of ethics and values systems.

Staying current on emerging research and practices

Analyse current research and practices in own professional field
Identify the impact of new research and technology on engineering practices.

Interpersonal Skills: Teamwork and Communication

Teamwork

Form Effective Teams

Identify the components of an effective team
Identify the stages of team formation
Identify team roles and their impact on team performance
Analyse the strengths and weakness of a team.

Manage and Participate in Teams

Identify goals and agenda

Apply team ground rules
Apply facilitation and conflict resolution strategies
Display teamwork, including leadership, in a range of team role situations.

Communications

Design appropriate communications strategies

Analyze the communication situation (e.g., in terms of purpose, audience and context (PAC))
Identify key considerations in communicating across cultures and disciplines
Identify communications objectives
Read critically and select relevant content
Identify and choose appropriate communication structure and style
Select appropriate multimedia and graphical communication (e.g., email, voicemail, video conferencing, tables and charts, sketching and drawing).

Demonstrate effective written communication

Write with logical organization and clear language flow
Use concise and precise language
Use correct grammar, spelling and punctuation
Apply appropriate written styles with appropriate formatting conventions to suit PAC.

Demonstrate effective oral communication

Design and deliver presentations applying communication design principles (e.g., as in 1.1.1 above)
Speak clearly and coherently (e.g., to be understood in a range of communication situations)
Use appropriate nonverbal communications (e.g., posture, gestures, eye contact)
Demonstrate active and empathetic listening in a range of communication situations (e.g., working in teams, responding to questions, etc.)
Ask and answer questions effectively.

Conceiving, Designing, Implementing and Operating Systems in the Enterprise and Societal Context

External and Societal Context

Understand Roles and Responsibility of Technologists

Explain professional goals and roles of the engineering profession
Analyse the responsibilities of technologists to society.

Understand the Impact of Engineering on Society

Explain the impact of engineering on the environment (e.g., ecological, social, economic, cultural systems etc.)
Explain the need for Sustainable Development
Identify possible solutions to support Sustainable Development.

Understand the Regulation of Engineering

Recognize the way in which legal systems regulate and influence engineering
Describe how professional societies license and set standards
Describe how intellectual property is created, utilized and defended.

Develop a Global Perspective

Identify the basis of cultural diversity
Compare and contrast a range of cultural practices and their impact on human conduct and communication
Define globalization
Identify factors that contribute to globalization
Identify the social, economic and environmental impact of globalization.

Enterprise and Business Context

Appreciate Different Enterprise Cultures

Recognize the differences in process, culture, and metrics of success in various enterprise cultures:
Corporate vs. academic vs. governmental vs. non-profit/NGO
Market vs. policy driven
Large vs. small
Centralized vs. distributed
Research and development vs. operations
Mature vs. growth phase vs. entrepreneurial
Longer vs. faster development cycles
With vs. without the participation of organized labor.

Identify Enterprise Strategy, Goals, and Planning

State the mission and scope of the enterprise
Recognize an enterprise's core competence and markets
Recognize the research and technology process
Recognize key alliances and supplier relations
List financial and managerial goals and metrics
Recognize financial planning and control
Describe stake-holder relations (with owners, employees, customers, etc.).

Understand Technical Entrepreneurship

Recognize entrepreneurial opportunities that can be addressed by technology
Recognize technologies that can create new products and systems
Describe entrepreneurial finance and organization.

Understand Organizational Structure and Dynamics

Define the function of management and organizational structure
Describe various roles and responsibilities in an organization
Describe the roles of functional and program organizations
Describe working effectively within hierarchy and organizations
Describe change, dynamics and evolution in organizations.

Conceiving and Engineering Systems

Identify market needs and opportunities

Elicit and interpret customer needs
Identify opportunities which derive from new technology or latent needs
Classify competitors and benchmarking information.

Define Function, Concept and Architecture

Identify necessary functions (and behavioral specifications)
Select relevant engineering concepts
Identify the appropriate level of technology
Analyse trade-offs and recombination of concepts
Identify high level architectural form and structure
Identify the decomposition of form into elements, assignment of function to elements, and definition of interfaces.

Model System to Verify Goals

Identify appropriate models and prototype, (e.g., models, animation tools, simulation etc.) of technical performance
Describe the concept of implementation and operations
Select between trade-offs among various goals, function, concept, structure and cost.

Develop Project plan

Produce project scope, objectives and deliverables
Identify project resources (e.g., cost, manpower, facilities, etc.)
Produce project schedule, key milestones and critical path analysis
Identify risks and contingency plans
Produce project report guidelines.

Designing

Formulate the Design

Choose requirements for each element or component derived from goals and requirements
Analyse alternatives in design
Evaluate prior work in the field, standardization and reuse of designs (including reverse engineer and redesign)
Select the most appropriate design
Synthesize the final design.

Plan the Design Process and Approaches

Explain the activities in the phases of system design (e.g., conceptual, preliminary, and detailed design)
Identify process models appropriate for particular development projects (waterfall, spiral, concurrent, etc.).

Apply Disciplinary Knowledge and Skills

Choose appropriate techniques, tools, and processes
Conduct quantitative analysis of alternatives
Evaluate analytical refinement of the design.

Apply a Multidisciplinary Perspective

Explain multidisciplinary design
Compare and contrast different discipline perspectives (e.g., conventions, assumptions, design environments)
Identify interactions between disciplines in design.

Evaluate design/prototype to achieve multiple objectives

Appraise design for:

- Performance, life cycle cost and value
- Optimizing experience through aesthetics and other human factors
- Implementation, verification, test and environmental sustainability
- Functionality
- Maintainability, reliability, and safety
- Robustness, evolution, product improvement and retirement.

Implementing

Plant the Implementation Process

State the goals for performance, cost and quality
Plan the implementation project according to:

Task allocation
Work flow.

Plan for Hardware Realization

Describe the production of parts
Describe the assembly of parts into larger constructs
Define tolerances, variability, key characteristics and statistical process control.

Planning for Software Implementing Process

Explain the breakdown of high level components into module designs (including algorithms and data structures)
Describe algorithms (data structures, control flow, data flow)
Describe the programming language
Execute the low-level design (coding)
Describe the system build.

Planning for Hardware Software Integration

Describe the integration of software in electronic hardware (size of processor, communications, etc.)
Describe the integration of software integration with sensor, actuators and mechanical hardware
Describe hardware/software function and safety.

Testing, Verifying, Validating, and Certifying

Describe test and analysis procedures (hardware vs. software, acceptance vs. qualification)
Describe the verification of performance to system requirements
Evaluate the validation of performance to customer needs
Verify design conformance to standards.

Managing Implementation

Identify sourcing, partnering, and supply chains
Control implementation cost, performance and schedule
Describe quality and safety assurance
Describe possible implementation process improvements.

Operating

Planning Training and Operation procedures

Interpret the goals for operational performance, cost, and value
Identify types training for professional operations (e.g., Simulation, demonstration, instructional procedures)

Plan a training program for end user
Produce operational process manual.

Managing Operations

Forming partnerships and alliances
Control operational cost, performance and scheduling
Describe quality and safety assurance
Describe possible operations process improvements.

Supporting the Product Lifecycle

Explain maintenance and logistics
Describe lifecycle performance and reliability
Recommend product improvement based on needs observed in operation
List disposal options
List environmental considerations for disposal.

About the Author

Dennis Sale is presently Senior Education Advisor at Singapore Polytechnic. He has previously worked across all sectors of the British educational system. Over the past 20 years Dennis has been extensively involved in training, coaching and assessing teaching professionals in a variety of vocational and cultural contexts. His specialist areas include Creative Teaching, Curriculum Development and Blended Learning. He has invented highly effective and practical models in these areas, provided a wide range of consultancies for both public and private sector organizations, and conducted numerous workshops in many countries. He is also widely noted to be a dynamic and creative presenter, blending practical relevance with situated humour.

D. Sale, *The Challenge of Reframing Engineering Education*,
DOI: 10.1007/978-981-4560-29-0, © Springer Science+Business Media Singapore 2014

FSC
www.fsc.org
MIX
Papier aus verantwortungsvollen Quellen
Paper from responsible sources
FSC® C105338

If you have any concerns about our products,
you can contact us on
ProductSafety@springernature.com

In case Publisher is established outside the EU,
the EU authorized representative is:
Springer Nature Customer Service Center GmbH
Europaplatz 3, 69115 Heidelberg, Germany

Printed by Libri Plureos GmbH
in Hamburg, Germany